Wave Propagation in Structures

James F. Doyle

Wave Propagation
in Structures

An FFT-Based Spectral
Analysis Methodology

Springer-Verlag
New York Berlin Heidelberg
London Paris Tokyo Hong Kong

James F. Doyle
School of Aeronautics and Astronautics
Purdue University
West Lafayette, Indiana 47907
USA

Library of Congress Cataloging-in-Publication Data
Doyle, James F., 1951-
Wave propagation in structures: an FFT-based spectal analysis
  methodology / James F. Doyle.
Bibliography: p. Includes index
ISBN 0-397-97078-9
1. Wave-motion, Theory of. 2. Spectral theory (Mathematics)
  3. Fourier transformations. I. Title.
QA935.D693   1989

Printed on acid-free paper.

Text provided by the author in camera-ready form.
Printed and bound by Edwards Brothers, Inc., Ann Arbor, Michigan.
Printed in the United States of America.

9 8 7 6 5 4 3 2 1

ISBN 0-387-97078-9 Springer-Verlag New York Berlin Heidelberg
ISBN 3-540-97078-9 Springer-Verlag Berlin Heidelberg New York

To my parents,
Patrick and Teresa Doyle,
on their forty-second wedding anniversary.

Preface

The study of wave propagation seems very remote to many engineers, even to those who are involved in structural dynamics. I think one of the reasons for this is that the examples usually taught in school were either so simple as to be inapplicable to real world problems, or so mathematically abstruse as to be intractable. This book contains an approach, spectral analysis, that I have found to be very effective in analyzing waves. What has struck me most about this approach is how I can use the same analytic framework to do predictions as well as to manipulate experimental data. As an experimentalist, I had found it very frustrating having my analytical tools incompatible with my experiments. For example, it is experimentally impossible to generate a step-function wave and yet that is the type of analytical solution available.

Spectral analysis is very encompassing — it touches on analysis, numerical methods, and experimental methods. I wanted this book to do justice to its versatility, so many subjects are introduced. As a result some areas may seem a little thin and I regret this. But I do hope, nonetheless, that the bigger picture, the unity, comes across.

To encourage you to try the spectral analysis approach I have included complete source code listings to some of the computer programs mentioned in the text. I do not have the space to provide the listings of all the programs, but they can be obtained from *ikayex* SOFTWARE TOOLS, 615 Elston Road, Lafayette, Indiana 47905, USA.

This book has been produced in camera-ready form using the PCTEX® implementation of the TEX and LATEX typesetting systems with a Cordata® LP300 laser printer. A very special thanks goes to Linda E. Brown for helping me learn these wonderful systems. She also typed all the difficult sections of the book with good grace and no complaints.

A book like this is impossible to complete without the help of many other people, but it is equally impossible to properly acknowledge them individually. However, I would like to single out Hong Zhang for checking much of the technical content. Any remaining errors and inaccuracies are purely my own doing.

March,1989   James F. Doyle

Contents

Introduction

"We must gather and group appearances, until the scientific imagination discerns their hidden laws, and unity arises from variety; and then from unity we must rededuce variety, and force the discovered law to utter its revelations of the future."

W. R. HAMILTON

The purpose of this small book is to introduce the spectral analysis method as a means of solving wave propagation problems in structures. The emphasis is on practical methods from both the computational and applications aspects, and reference to experimental results is made whenever possible.

While it is possible to solve structural dynamics problems by starting with the partial differential equations of motion and integrating, the task is horrendously large even for the biggest computers available. This would not be a good idea anyway because (and this is a point very often overlooked) a true solution to a problem is one that also puts organization and coherence onto the results. It is not sufficient to be able to quote, say, the strain history at some location or even at thousands of locations. The results must be placed in some higher-order context, be seen as part of some larger unity. Notice that this aspect of the problem is present even when interpreting experimental results and is not just associated with analysis. One of the goals of this book is to provide such a unifying framework for the analysis of waves in structures. By consistently using the spectral analysis method for all problems a unity emerges. This unity is not only in the formulation but (when coupled with the fast Fourier transform) is also among the formulation, the solution procedure, the solutions themselves, and the post-manipulations of the results.

What Is Spectral Analysis?

Over the years many analytical techniques have been developed for treating wave propagation problems. Central among these is the method of Fourier synthesis (or spectral analysis), where the behavior of a signal is viewed as a superposition of many infinitely long wave trains of different periods (or frequencies). The actual response is synthesized by a judicious combination of these wave trains. Thus the problem of characterizing a signal is transformed into one of determining the set of combination coefficients. These coefficients are called the Fourier *transform* of the signal. While the problem being tackled invariably simplifies when it is expressed in terms of the Fourier transform (Sneddon* shows a wide range of applications to both static and dynamic problems) the last step in the analysis involves performing an inverse transform (reconstructing the signal) and this, generally, is very difficult to do in an exact manner. Consequently, many approximate and asymptotic schemes have usually been resorted to. These are quite adequate for determining the remote behavior (as is required in seismology, say) but can lose too much information when applied to structural impact problems. It should also be pointed out that analytical transforms are feasible only if the function to be transformed is mathematically simple — unfortunately this is not the case in any situation of practical interest and is certainly not true when dealing with experimental data. This inversion problem is the biggest impediment to a more widespread use of the transform methods.

Spectral analysis forms the basis of this book, but the approach is different from the classical method in that, from the outset, the transforms are approximated by the discrete Fourier transform (DFT). In contrast to the continuous transform, this represents the signal by a finite number of wave trains and has the enormous advantage in that the fast Fourier transform (FFT) computer algorithm can be used for economically computing the transforms. Being able to do transforms and inversions quickly adds great heuristic value to the tool in that the waves can be actually "seen" and iterated on, and realistic signals (even ones that are experimentally based) can be treated. It should be pointed out that while the method uses a computer, it is not a numerical method in the usual sense, because the analytical description of the waves is still retained. As a consequence, the very important class of inverse problems can be tackled.

The approach presented in this book takes advantage of many of the techniques already developed for use in time series analysis and for the efficient numerical implementation of them. In fact, this aspect of spectral analysis is really part of the more general area of digital processing of the signals and herein lies one of the unifying advantages of the present approach — the programming structure is then already in place for 'the subsequent post-processing of the data. This is especially significant for the manipulation of experimental data. Chatfield's book has served as a readable introduction to the area.

Structures as Waveguides

A structure can be as simple as a cantilevered diving board or as complicated as the proposed space platforms that are three-dimensional multimember jointed grids with many attachments. To make headway with a structural dynamics problem,

---

* Introductory references are collected on page 6.

it is necessary to approximate it as a collection of waveguides with appropriate connectivities at the joints. A waveguide directs the wave energy along its length and, in its elementary form, can be viewed as a hydraulic or electric network analog. Unfortunately, the quantities transported are more complicated than water flow or current.

Perceiving a structure as a waveguide is not always an easy matter. For example, it is reasonable to expect that a narrow rod struck along its length will conduct longitudinal waves and intuition says that if it is struck transversely it generates flexural motion. However, on closer examination it turns out not to be that simple. When the beam is first impacted on the side, the waves generated propagate into a semi-infinite body and behave as if there is only one surface. Only after some time has elapsed do the waves experience the other lateral surface, where they then reflect back into the body. On a time scale comparable to many transits of the wave, it is seen that a particle at some location further down the guide experiences a complex superposition of the initial wave plus all the new ones generated by reflections. This obviously is not flexural motion. The question then is: At what stage (both time and position) does the response resemble a flexural wave, if at all?

The answer comes in two parts. First, it can be said that the transition depends on such factors as the duration of the pulse, the distance between bounding surfaces, and the transit time. That is, the larger the pulse and the smaller the depth, the sooner (both in time and position) the response resembles a flexural wave. However, it never does become a flexural wave. This leads to the other part of the answer. The point and success of waveguide analysis is to forego a detailed analysis of the waves and replace the three-dimensional model by a simpler one that has embedded in it the essential characteristics of the behavior as well as a reasonable approximation of the lateral boundary conditions. This model usually involves resultants on the cross-section and is valid (within itself) for all time and positions not just at large times and distances.

There are various schemes for establishing the waveguide model that range from the purely *ad hoc* "Strength of Materials" approaches, to reduced forms of the 3-D equations, to using exact solutions. Redwood gives a good survey of waveguide analysis for both solids and fluids. The approach taken in this book is to begin the analysis using an elementary model and then to add complexity to it. This has the advantage of retaining whatever intuitive aspects are there; in addition, it can show the way to approach as yet unformulated problems.

Wave Propagation and Vibrations

A major aspect of this book is the persistent treatment of the effect of boundaries and discontinuities on the waves because real structures have many such terminations. This can be done efficiently because the quantities used in the analysis of the waveguide are also used to set up the connectivity conditions. As a result, wave solutions for structures more interesting than simple rods and beams can be pieced together successfully. Moreover, the way to solve problems of structures with many members and boundaries is then available.

A connection not well understood by those studying structural dynamics is the relationship between wave propagation and vibrations. For many engineers, these are two separate areas with quite distinct methods of analysis. However, another advantage of the spectral approach to dynamics is that the close connection between waves and vibrations becomes apparent. Even the same language can be used, terms such as power spectral density, filtering, spectral estimation, convolution, and sampled waveforms have the same meaning. Consequently, many of the technologies developed over the last twenty years for vibrations and modal analysis are directly applicable to the spectral analysis of waves.

An exciting possibility (and one of the motivations for writing this book) is to facilitate the reverse process, that is, to transport many of the waves ideas into vibration analysis. This should lead to a richer understanding of such topics as impulse testing and transient vibrations.

Outline of the Book

This book concentrates on wave propagation in the basic structural elements of rods, beams, and plates. These form a rich collection of problems and the intent is to show that they all can be analyzed within the same framework once the spectral analysis approach is adopted. Because of the structural applications (and since all structures are finite in extent), emphasis is placed on the interaction of the wave with discontinuities such as boundaries, junctions, and attachments. While the material of each chapter is reasonably self-contained, Graff's book can be used as an excellent supplemental reference. The book by Elmore and Heald gives a broader and simpler introduction to waves.

Chapter 1 recapitulates the essence of the continuous Fourier transform and its approximation in the form of the discrete Fourier series. The factors affecting the quality of the approximation (or spectral estimate) are elucidated. The second chapter discusses, in a general way, how spectral analysis can be used to solve differential equations and especially those associated with wave motion. It is shown how the frequency acts as a scale on the time domain while the wavenumber acts as a scale on the space domain. Two concepts of significance emerge from this. One is the idea of the spectrum relation (that unique relation between the frequency and the wavenumber) and is essentially the transform equivalent of the space-time behavior. The other is that of multi-mode solutions. These are shown to play a fundamental role in the solution of actual boundary value problems even though all are not necessarily propagating modes.

The following two chapters deal, respectively, with rods and beams, in nearly the same format. First, the governing differential equations are derived and then, by spectral analysis, the kernel solutions and spectrum relations are obtained. Having initiated a wave, how it interacts with structural discontinuities is then investigated. The presented framework of spectral analysis can also conveniently handle higher-order structural theories for wave propagation without additional complexity. These chapters present the higher mode theories of Mindlin-Herrmann for rods and Timoshenko for beams, for example. A foundation in these is provided that will readily allow extensions to be made to other specialized problems. Finally, each chapter ends by introducing extensions to the analysis necessary to treat such aspects as viscoelasticity or variable dimensions.

Chapter 5 introduces a matrix method approach to structural dynamics that combines the finite element method and the spectral analysis method. Its application to wave motion in connected structures is demonstrated. This can be viewed as an extension of the analysis to multiple dimensions and clearly makes the study of wave propagation in complicated structures practical. A different multi-dimensional extension is shown in Chapter 6 when dealing with plates. Here the implications of double series and infinite modes are investigated.

The spectral methodology is an ideal companion for experimental analysis, and throughout the chapters (especially the last one) summaries of some of its applications are given. The emphasis is on how spectral analysis can extend the type of information extracted from the experimental data — for example, how a structural response can be used to infer the forcing history, or how measurements can be used to determine the dynamic material properties over a large frequency range.

Admittedly, a large range of problems (such as those associated with scattering and diffraction) have been left out even though most of them are treatable by the spectral methods. Consequently an effort is made to supplement each chapter with a collection of pertinent references that indicate extensions of the theory and the applications. These can form the basis for further study.

Wave Analysis on a PC

This book attempts to be a guide to the practical analysis of waves in structures and so the source code to many computer programs used in doing spectral analysis are provided in the Appendix. Of particular value are the FFT program CFFTCOMP and the rod program RODCOMP. This latter program was used to run all the examples of Chapter 3 and can serve as the blueprint for similar programs dealing with beams and plates. Space constraints forbid providing a user's guide, so the programs have been written menu-style and are straightforward to run. A complete collection of programs is available through the author.

All of the programs provided are written in Microsoft® FORTRAN 4.0 for the IBM®-AT and compatibles family of personal computers. Since most of the code is compatible with the ANSI 77 standard (as described by Balfour), then there should be no difficulties in porting them to other systems. They run quite fast without using any special features. Consequently, the user can interact with the analysis and the solution quite unlike any other method. This also is part of the unity referred to at the beginning; the solution procedure and the analysis of the results are one.

Introductory References

1. Balfour, A. and Marwick, D.H., 1979, *Programming in Standard FORTRAN 77*, North-Holland, Inc., New York.

2. Chatfield, C., 1984, *The Analysis of Time Series: An Introduction*, Chapman and Hall, London.

3. Elmore, W.C. and Heald, M.A., 1985, *Physics of Waves*, Dover, New York.

4. Graff, K.F., 1975, *Wave Motion in Elastic Solids*, Ohio State University Press.

5. Redwood, M., 1960, *Mechanical Waveguides*, Pergamon Press, New York.

6. Sneddon, I.N., 1951, *Fourier Transforms*, McGraw-Hill, New York.

Chapter 1

Spectral Analysis and the FFT

It has long been known that an arbitrary time signal can be thought of as the superposition of many sinusoidal components. This is the basis of Fourier (or spectral) analysis. In wave analysis, the time domain for the disturbance is from minus infinity to plus infinity and thus the components have a continuous distribution (known as the continuous Fourier transform). However, the numerical evaluation of the transform requires discretizing the distribution in some manner, and the one chosen here is by way of the discrete Fourier transform (DFT.) This has two advantages. First, many of the ideas and methods of time series analysis [33] can be borrowed and used for the present purposes. Second, it allows the use of the very efficient fast Fourier transform (FFT) computer algorithm.

The goal of this chapter, then, is to introduce the fast Fourier transform numerical algorithm for the efficient computation of the spectral content of a signal. Possible sources of errors in using it on finite samples (such as leakage and aliasing) and ways of reducing their influences will be discussed. Background information on signal processing can be found in References [20,23,33].

1.1 Continuous Fourier Transforms

The continuous transform is a convenient starting point for discussing spectral analysis because of its exact representation of functions. Only the basic definition of the continuous transform is given here, a fuller account can be found in References [179,205].

Continuous Transform

The continuous Fourier transform pair of a function $F(t)$, defined on the time domain from $-\infty$ to $+\infty$, is given as:

$$2\pi F(t) = \int_{-\infty}^{\infty} C(\omega)e^{+i\omega t}d\omega , \qquad C(\omega) = \int_{-\infty}^{\infty} F(t)e^{-i\omega t}dt \qquad (1.1)$$

where $C(\omega)$ is the continuous Fourier transform (CFT), ω is the angular frequency, and i is the complex $\sqrt{-1}$. (In general, both the time domain and frequency domain functions are complex.) The first form is the inverse transform while the second is the forward transform — this arbitrary convention arises because the signal to be transformed usually originates in the time domain. The factor of 2π is necessary so that a sequential use of the forward and inverse transforms recovers the original function. However, it should be pointed out that other definitions for this factor can be found in the literature. The full symmetry of the transforms can be seen by introducing the cyclic frequency $2\pi f = \omega$ and rearranging the inverse relation as

$$F(t) = [\int_{-\infty}^{\infty} C^*(f)e^{-i2\pi ft}df]^*, \qquad C(f) = \int_{-\infty}^{\infty} F(t)e^{-i2\pi ft}dt \qquad (1.2)$$

where the superscript * refers to the complex conjugate. This form will be utilized later in the numerical implementation because it allows a single algorithm to be written for both the forward and inverse transforms. It is a matter of taste as to which frequency, ω or f, is used — it is convenient that both of them be used in this work.

The process of obtaining the Fourier transform of a signal separates the waveform into its constituent sinusoids (or spectrum) and thus a plot of $C(\omega)$ against frequency represents a diagram displaying the amplitude of each of the constituent sinusoids. The spectrum $C(\omega)$ has both real and imaginary parts.

Example of a Rectangular Pulse

By way of a simple example of the application of Fourier transforms, consider a rectangular pulse where the time function is given by

$$\begin{aligned} F(t) &= F_o & -a/2 \leq t \leq a/2 \\ &= 0 & \text{otherwise} \end{aligned}$$

Substituting into the continuous transform of equation (1.1) gives

$$2\pi C(\omega) = 2F_o \left\{ \frac{\sin(\omega a/2)}{\omega} \right\} = F_o a \left\{ \frac{\sin(\omega a/2)}{\omega a/2} \right\} \qquad (1.3)$$

In this particular case the transform is real-only and symmetric about $\omega = 0$ as shown in Figure 1.1. The term inside the braces is the *sinc* function, and has the characteristic behavior of starting at unity magnitude and oscillating with decreasing amplitude. Note that the value of the transform at $\omega = 0$ is just the area under the time function.

When the pulse is displaced along the time axis such that the function is given by

$$\begin{aligned} F(t) &= F_o & t_o \leq t \leq t_o + a \\ &= 0 & \text{otherwise} \end{aligned}$$

the transform is then

$$2\pi C(\omega) = F_o a \left\{ \frac{\sin(\omega a/2)}{\omega a/2} \right\} e^{-i\omega(t_o + a/2)} \tag{1.4}$$

which has both real and imaginary parts and is not symmetric. On closer inspection, however, it is seen that the magnitudes of the two transforms are the same, it is just that the latter is given an extra phase shift of amount $\omega(t_o + a/2)$. Figure 1.1 shows the transform for different amounts of shift.

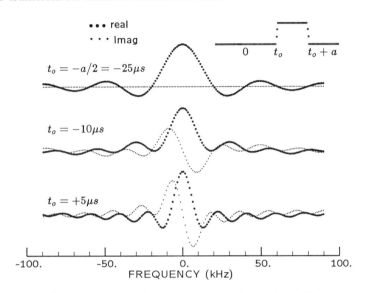

Figure 1.1: Continuous transform of a rectangular pulse with $a = 50\mu s$.

If a pulse is visualized to be at different positions relative to the origin, then the amplitude of the spectra will be the same, but each will have a different phase shift. That is, movement in the time domain causes phase shifts in the frequency domain. Investigating these phase shifts is the fundamental application of spectral analysis to wave propagation and will be pursued in Chapter 2. The remainder of this chapter deals with practical ways of estimating the spectral content of a signal.

1.2   Properties of Fourier Transforms

For completeness, this section summarizes some of the major properties of Fourier transforms; more detailed accounts can be found in References [179,207]. In all cases, the results can be confirmed by taking the example of the rectangular pulse and working through the transforms long hand. (This is a useful exercise, in any event, to become familiar with manipulating these transforms.) To aid in the summary the transform pair

$$F(t) \leftrightarrow C(\omega)$$

will be referred to, where the symbol $\leftrightarrow$ means "can be transformed into". The double arrowheads reinforce the idea that the transform can go in either direction and that its properties are symmetric.

Linearity

If the functions $F_A(t)$ and $F_B(t)$ have the transforms $C_A(\omega)$ and $C_B(\omega)$, respectively, then the combined function $[F_A(t)+F_B(t)]$ has the transform $[C_A(\omega)+C_B(\omega)]$. That is

$$F_A(t) + F_B(t) \leftrightarrow C_A(\omega) + C_B(\omega) \qquad (1.5)$$

This is an essential property of the transform and it means that if a signal is composed of the simple sum of two contributions (say an incident wave and a reflected wave) then the transform is also composed of a simple sum of the separate transforms. This linearity property is at the heart of superposition.

Scaling

The function $F(kt)$ (where k is a non-zero constant) has the transform pair

$$F(kt) \leftrightarrow \frac{1}{|k|} C(\omega/k) \qquad (1.6)$$

Therefore, time domain compression corresponds to frequency domain expansion (and vice versa.) For example, if the width of a pulse is narrow in the time domain, then its extent in the frequency domain is broad. It should also be noted that the amplitude decreases because the energy is distributed over a greater range of frequencies.

Time Shifting

If the function $F(t)$ is time shifted by t_o, then it has the transform pair

$$F(t - t_o) \leftrightarrow C(\omega)e^{-i\omega t_o} \qquad (1.7)$$

This property was already seen in the example of the last section where the rectangle was displaced from the origin. Of course, the property refers to any change at any position. There is a corresponding relation for frequency shifting.

Even and Odd Functions

Any general function of time can be written as the sum of an even and odd function about $t = 0$. That is, letting $F(t) = F_e(t) + F_o(t)$ gives

$$F_e(t) = \tfrac{1}{2}[F(t) + F(-t)], \qquad F_o(t) = \tfrac{1}{2}[F(t) - F(-t)]$$

By use of linearity and scaling these have the transform pairs

$$F_e(t) \quad \leftrightarrow \quad \tfrac{1}{2}[C(\omega) + C(-\omega)] = \text{Real}[C(\omega)]$$
$$F_o(t) \quad \leftrightarrow \quad \tfrac{1}{2}[C(\omega) - C(-\omega)] = i\text{Imag}[C(\omega)] \tag{1.8}$$

The real part of the transform gives the even part of the time function, while the imaginary part gives the odd function.

Since a pulse propagating in time is inherently neither even nor odd, then the transforms encountered later will always be complex.

Real Time Functions

The transform pairs above are valid for both $F(t)$ and $C(\omega)$ being complex, but the functions of usual interest in wave analysis are when $F(t)$ is real. To see the effect of this, rewrite equation (1.1) in terms of its real and imaginary parts as

$$2\pi F_R = \int [C_R \cos \omega t - C_I \sin \omega t] d\omega, \qquad C_R = \int [F_R \cos \omega t + F_I \sin \omega t] dt$$

$$2\pi F_I = \int [C_R \sin \omega t + C_I \cos \omega t] d\omega, \qquad C_I = -\int [F_R \sin \omega t - F_I \cos \omega t] dt$$

where the following decomposition is used

$$\cos \theta = \tfrac{1}{2}[e^{i\theta} + e^{-i\theta}], \qquad \sin \theta = -i\tfrac{1}{2}[e^{i\theta} - e^{-i\theta}] \tag{1.9}$$

From the above it is apparent that when $F(t)$ is real-only, that C_R is even and C_I is odd. Mathematically, this expressed as

$$C_R(-\omega) = C_R(\omega), \qquad C_I(-\omega) = -C_I(\omega)$$

which says that the functions are symmetrical and antisymmetrical, respectively, about the zero frequency point. This can also be expressed as saying that the negative frequency side of the transform is the complex conjugate of the positive side and is written as

$$C(-\omega) = C^*(\omega) \tag{1.10}$$

Advantage will be taken of this property when setting up the algorithm to obtain the transform of real signals.

Convolution

A very interesting property arises in connection with the products of functions. Consider the transform of two time functions

$$C(\omega) = \int F_A(t) F_B(t) e^{-i\omega t} dt$$

Using the inverse transform relation for F_A gives

$$C(\omega) = \iint C_A(\bar{\omega}) e^{+i\bar{\omega}t} d\bar{\omega} \, F_B(t) e^{-i\omega t} \, dt$$

This can be further rearranged as

$$C(\omega) = \int C_A(\bar{\omega}) \int F_B(t) e^{-i(\omega - \bar{\omega})t} \, dt \, d\bar{\omega} = \int C_A(\bar{\omega}) C_B(\omega - \bar{\omega}) \, d\bar{\omega}$$

and is expressed as the transform pair

$$F_A(t) F_B(t) \Leftrightarrow \int C_A(\bar{\omega}) C_B(\omega - \bar{\omega}) d\bar{\omega} \tag{1.11}$$

This particular form is called a *convolution*. It will have use in understanding the effects of sampling and filtering on the computed transforms. For example, a signal truncated in the time domain can be thought of as the product of the original signal with the truncating signal. The result in the frequency domain is no longer a simple representation of either. This is considered in more detail later in this chapter.

There is a similar relation for products in the frequency domain, namely,

$$C_A(\omega) C_B(\omega) \Leftrightarrow \int F_A(\tau) F_B(t - \tau) \, d\tau \tag{1.12}$$

This shows that a time domain convolution can be performed as frequency domain multiplication. Even though this involves both a forward and inverse transform it is computationally faster than performing the convolution directly. All the mechanical systems considered in the later chapters will be represented in the frequency domain as products of the input times the system response. In fact, the reason why the frequency domain is so useful for performing computations is because when viewed (or treated) in the time domain they are complicated convolution integrals.

1.3 Fourier Series

The continuous Fourier transform is a powerful technique but has the drawback that the functions (signals) must be known analytically. (That is, be a square wave, or step function, or the like.) This occurs in only rare cases making it unsuitable for practical situations, especially if the signals are experimental in origin. Before considering the discrete Fourier transform (DFT), it is instructive to first look at Fourier series since, in a sense, this is half-way between the two.

A Fourier series representation of a function assumes the function is periodic in the time domain. Consequently, it can only be an approximation to the transient functions found in wave analysis since these are assumed to be present on the full domain. But it will be shown that under the proper circumstances the approximation can be very accurate. In elementary terms, if the signal has a large duration of zero amplitude and the analysis assumes it repeats itself on a period very large compared to the time of interest of the wave, then for all intents and purposes it behaves as an infinite wave.

Basic Representation

A function $F(t)$ with period T can be expressed in the form of a Fourier series as

$$F(t) = \frac{1}{2}a_o + \sum_{n=1}^{\infty}[a_n \cos(2\pi n\frac{t}{T}) + b_n \sin(2\pi n\frac{t}{T})] \tag{1.13}$$

where the coefficients are obtained from

$$a_n = \frac{2}{T}\int_0^T F(t)\cos(2\pi n\frac{t}{T})dt, \qquad n = 0, 1, 2, \ldots$$
$$b_n = \frac{2}{T}\int_0^T F(t)\sin(2\pi n\frac{t}{T})dt, \qquad n = 1, 2, \ldots \tag{1.14}$$

If n ranges over both positive and negative numbers, then it is apparent that a_n is symmetric while b_n is antisymmetric about $n = 0$. Using the identities of equation (1.9) and introducing the discrete frequency as $\omega_n = 2\pi n/T$, the above can be written as

$$F(t) = \frac{1}{2}a_o + \frac{1}{2}\sum_{n=1}^{\infty}(a_n - ib_n)e^{i\omega_n t} + \frac{1}{2}\sum_{n=1}^{\infty}(a_n + ib_n)e^{-i\omega_n t}$$
$$a_n + ib_n = \frac{2}{T}\int_0^T F(t)e^{-i\omega_n t}\,dt, \qquad n = 0, 1, 2, \ldots$$
$$a_n - ib_n = \frac{2}{T}\int_0^T F(t)e^{-i\omega_n t}\,dt, \qquad n = 0, 1, 2, \ldots$$

Since the phase $\omega_n t$ is linear in n, then $F(t)$ can be further written as

$$F(t) = \frac{1}{2}\sum_{n=-\infty}^{-1}(a_{-n} + ib_{-n})e^{i\omega_n t} + \frac{1}{2}a_o + \frac{1}{2}\sum_{n=1}^{\infty}(a_n - ib_n)e^{i\omega_n t}$$

This permits the simpler, more compact, representation

$$F(t) = \frac{1}{2}\sum_{-\infty}^{\infty}(a_n - ib_n)e^{i\omega_n t} = \sum_{-\infty}^{\infty}C_n e^{i\omega_n t} \tag{1.15}$$

where

$$C_n = \tfrac{1}{2}(a_n - ib_n) = \frac{1}{T}\int_0^T F(t)e^{-iw_n t}\, dt\,, \qquad n = 0, \pm1, \pm2\ldots \qquad (1.16)$$

From the above form, it is apparent that

$$a_{-n} = a_n\,, \qquad b_{-n} = -b_n\,, \qquad n = 1, 2, 3, \ldots$$

which is the Fourier series result equivalent to equation (1.10).

The signal repeats itself every T (seconds), therefore, it has a base, or fundamental, frequency of $w_o/2\pi = f_o = 1/T$ (hertz). The time signal is represented in terms of this fundamental frequency as

$$F(t) = \sum_{n=-\infty}^{\infty} C_n e^{i2\pi n f_o t} = \sum_{n=-\infty}^{\infty} C_n e^{in w_o t} \qquad (1.17)$$

In the previous representations the frequencies ranged from $-\infty$ to ∞, but from the above it is apparent that the "negative frequencies" are to be interpreted solely in terms of the range of n in the exponent and not physically.

Example of a Rectangular Pulse

Consider the transform of the rectangular pulse of the last section. The coefficients are given by

$$
\begin{aligned}
C_n &= \frac{1}{T}\int_{t_o}^{t_o+a} F_o e^{-i2\pi n t/T}\, dt = \frac{F_o}{T}\left[\frac{e^{-i2\pi n t/T}}{-i2\pi n t/T}\right]_{t_o}^{t_o+a} \\
&= \frac{F_o a}{T}\left\{\frac{\sin(\pi n a/T)}{\pi n a/T}\right\} e^{-i(t_o+a/2)2\pi n/T} \qquad (1.18)
\end{aligned}
$$

This result is the product of three terms. The first is the size of the pulse averaged over the period. The third is a phase shift due to the shifting of the pulse relative to the origin. For example, if the pulse is symmetric about $t = 0$, then $t_o = -a/2$ and there is no phase shift of the transform. The second term is the core of the transform and its amplitude is shown plotted in Figure 1.2 for various periods. Notice that the spacing of the coefficients is at every $1/T$ hertz in the frequency domain. For the periods $T = 500, 200, 100\,\mu s$ this gives spacings of $\Delta f = 2, 5, 10$ hertz, respectively. Therefore, for a given pulse, a larger period gives a more dense distribution of coefficients, approaching a continuous distribution in the limit of an infinite period. This, of course, is the continuous transform limit.

Let the width of the pulse decrease relative to the period. The obvious consequence is that there are more coefficients before the first zero. When both results are plotted against the argument $\pi n a/T$, then it is seen that the narrower pulse causes a greater density of coefficients.

Notice that in all the examples of Figure 1.2, the Fourier series gives the exact values of the continuous transform (except for a scale factor), but it does so only at discrete frequencies.

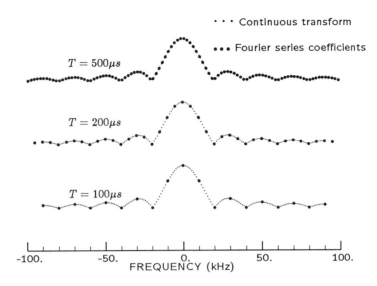

Figure 1.2: Fourier series coefficients for a rectangular pulse of duration $a = 50\mu s$.

Relationship to the Continuous Transforms

Much of the literature on wave propagation use Fourier integral or continuous transforms. It is of interest, therefore, to show the relationship between it and Fourier series.

First, the series transform pair is rewritten as

$$2\pi F(t) = \sum_{-\infty}^{\infty}(TC_n)e^{i\omega_n t}\Delta\omega\,, \qquad TC_n = \int_0^T F(t)e^{-i\omega_n t}dt$$

where $\omega_n = 2\pi n/T$ and $\Delta\omega = 2\pi/T$. Now let the time window encompass from $-\infty$ to $+\infty$ (assuming the signal is zero before time $t = 0$), thus $\Delta\omega$ becomes $d\omega$ and the summation becomes an integral. That is

$$F(t) \Rightarrow \int_{-\infty}^{\infty} TC_n(\omega)e^{i\omega t}d\omega\,, \qquad TC_n(\omega) \Rightarrow \int_{-\infty}^{\infty} F(t)e^{-i\omega t}\,dt$$

These are the Fourier continuous transforms. In the region of applicability, they have the correspondences

$$TC_n \Rightarrow C(\omega) \qquad\qquad (1.19)$$

An alternative scheme to set up the correspondence is to consider the C_o component for the rectangular pulse and how it relates to $C(0)$ of the continuous transform.

It can be shown quite simply that the Fourier series enjoys all the same properties of the continuous Fourier transform. The only significant difference is that for the Fourier series the time domain function must be periodic.

1.4   Discrete Fourier Transform

The discrete coefficients in the Fourier series are obtained by performing continuous integrations over the time period. These integrations are now replaced by summations as a further step in the numerical implementation of the continuous transform. In the process, however, some significant points arise, and it is important to be aware of them.

Approximation for the Fourier Coefficients

The essential starting point for the discrete transform is the series transform pair rearranged as

$$f(t) = \frac{1}{T} \sum_{-\infty}^{\infty} C_n e^{i\omega_n t}, \qquad C_n(\omega) = \int_0^T f(t) e^{-i\omega_n t}\, dt \tag{1.20}$$

In reference to Figure 1.3, let the function $f(t)$ be divided in M, piecewise constant, segments whose heights are f_m and base $\Delta T = T/M$. The coefficients are now obtained from

$$\begin{aligned}
C_n \approx D_n &= \sum_{m=0}^{M-1} f_m \int_{t_m - \Delta T/2}^{t_m + \Delta T/2} e^{-i\omega_n t}\, dt = \Delta T \sum_m f_m \left\{ \frac{\sin \omega_n \Delta T/2}{\omega_n \Delta T/2} \right\} e^{-i\omega_n t_m} \\
&= \Delta T \left\{ \frac{\sin \omega_n \Delta T/2}{\omega_n \Delta T/2} \right\} \sum_m f_m\, e^{-i\omega_n t_m}
\end{aligned} \tag{1.21}$$

It is seen that this is the sum of the transforms of a series of rectangles each shifted in time by $t = t_m + \Delta T/2$. The contribution of each of these will now be evaluated more closely.

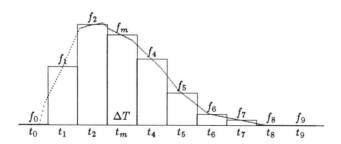

Figure 1.3: Discretization scheme for an arbitrary time function.

First look at the summation term. If $n > M$, that is, if $n = M + n^*$, then the exponential term becomes

$$e^{-i\omega_n t_m} = e^{-in\omega_o t_m} = e^{-iM\omega_o t_m} e^{-in^* \omega_o t_m} = e^{-i2\pi m} e^{-in^* \omega_o t_m} = e^{-in^* \omega_o t_m}$$

Hence the summation contribution simply becomes

$$\sum_{m=0}^{M-1} f_m e^{-in^* \omega_o t_m}$$

showing that it evaluates the same as when $n = n^*$. More specifically, if $M = 8$, then $n = 9, 11, 17$ evaluate the same as $n = 1, 3, 1$, respectively. The discretization process has forced a periodicity into the frequency description.

Looking at the other contribution, it is seen that the *sinc* function term does depend on the value of n and is given by

$$\text{sinc}(x) \equiv \frac{\sin(x)}{x}, \qquad x = \omega_n \Delta T / 2 = \pi n f_0 \Delta T = \pi n \frac{1}{T} \frac{T}{M} = \pi \frac{n}{M}$$

Consequently, the first zero occurs where $n = M$. This function is such that it decreases rapidly with n and is negligible beyond its first zero at $n = M$. If M is made very large, that is, the integration segments are very small, then it will be the higher order coefficients (i.e., large n) that are in the vicinity of the first zero. Let it be further assumed that the magnitude of these higher order coefficients is negligibly small. Then an approximation for the coefficients is

$$D_n \approx \Delta T \{1\} \sum_{m}^{M} f_m e^{-i\omega_n t_m}$$

on the assumption that it is good for $n < M$ and that $C_n \approx 0$ for $n \geq M$.

Since there is no point in evaluating the coefficients for $n > M - 1$, then the approximation for the Fourier series coefficients (now called the discrete Fourier transform) is taken as

$$f_m = f(t_m) \approx \frac{1}{T} \sum_{n=0}^{N-1} D_n e^{+i\omega_n t_m} = \frac{1}{T} \sum_{n=0}^{N-1} D_n e^{+i2\pi nm/N}$$

$$D_n = D(\omega_n) \approx \Delta T \sum_{m=0}^{N-1} f_m e^{-i\omega_n t_m} = \Delta T \sum_{m=0}^{N-1} f_m e^{-i2\pi nm/N} \qquad (1.22)$$

where both m and n range from 0 to $N - 1$. These are the definitions of the discrete Fourier transform. It is interesting to note that the exponentials do not contain dimensional quantities, only the integers n, m, N. In this transform, both the time and frequency domains are discretized, and, as a consequence, the transform be haves periodically in both domains.

The dimensional scale factors ΔT, $1/T$ are retained so that the discrete transform gives the same numerical values as the continuous transform. There are other possibilities for these scales found in the literature.

Properties of the Discrete Transform

A correspondence between the continuous and the discrete Fourier transform can be shown. First note the correspondence between time and frequency

$$t \Rightarrow t_m = m\Delta T, \qquad \omega \Rightarrow \omega_n = 2\pi n/(N\Delta T)$$

Also $f_m = f(t_m)$ is the input signal for the discrete Fourier transform, while $f(t)$ is the input signal for the continuous transform. Thus

$$C \leftrightarrow D_n \tag{1.23}$$

That is, they both have the same magnitudes. This was achieved by choosing the appropriate scale factors of equation (1.20).

Again, it can be shown quite simply that the discrete transform enjoys all the same properties of the integral transform. The only significant difference is that for the discrete transform both the time domain and frequency domain functions are now periodic. To put this point into perspective, consider the following: A discrete Fourier transform seeks to represent a signal (known over a finite time T) by a finite number of frequencies. Thus it is the continuous Fourier transform of a periodic signal. Alternatively, the continuous Fourier transform itself can be viewed as a discrete Fourier transform of a signal with an infinite period. The lesson is that by choosing a large signal sample length, the effect due to the periodicity assumption can be minimized and the discrete Fourier transform approaches the continuous Fourier transform.

Numerical Example

To help better illustrate the properties of the discrete transform consider the following numerical example. This can also serve as the test case for the program CFFTCOMP given in Appendix D.

Consider a real-only function given by the following sampled values

$$f_1 = f_2 = 1, \qquad f_0 = f_3 = f_4 = f_5 = f_6 = f_7 = 0$$

with $\Delta T = 1, N = 8$. This has the shape of a rectangular pulse. Eight points are represented; thus, it is implicit that the signal repeats itself beyond that. That is, the next few values are 0, 1, 1, 0, ... The transform becomes

$$D_n = \sum_{m=0}^{7} f_m e^{-i2\pi nm/8} = f_1 e^{-i\pi n/4} + f_2 e^{-i\pi n/2}$$

This gives, in explicit form, for the first ten transform points

$$D_0 = 2.0$$

$$D_1 = 0.707 - 1.707i$$
$$D_2 = -1.0 - 1.0i$$
$$D_3 = -0.707 + 0.293i$$
$$D_4 = 0.0$$
$$D_5 = -0.707 - 0.293i$$
$$D_6 = -1.0 + 1.0i$$
$$D_7 = 0.707 + 1.707i$$
$$D_8 = 2.0$$
$$D_9 = 0.707 - 1.707i$$

The obvious features of the transform are that it is complex and that it begins to repeat itself beyond $n = 7$. Note also that D_4 (the $(N/2 + 1)^{th}$ value) is called the Nyquist value. The real part is symmetric about the Nyquist, while the imaginary is anti-symmetric. From this it follows that the sum $\frac{1}{2}[D(n) + D(N - n)]$ gives only the real part, i.e.,

$$2.0, \quad 0.707, \quad -1.0, \quad -0.707, \quad 0.0, \quad -0.707, \quad \cdots$$

while the difference $\frac{1}{2}[D(n) - D(N - n)]$ gives the imaginary sequence

$$0, \quad -1.707i, \quad -i, \quad +0.293i, \quad 0, \quad -0.293i, \quad \cdots$$

These two functions are the even and odd decompositions, respectively, of the transform.

It is clear that if the input signal is complex only (for example, $f_1 = f_2 = i$), then the real part has the sequence

$$0, \quad 1.707, \quad 1.0, \quad -0.293, \quad 0, \quad +0.293, \quad \cdots$$

This corresponds to the odd decomposition. The imaginary part has

$$2.0i, \quad 0.707i, \quad -1.0i, \quad -0.707i, \quad 0.0, \quad -0.707i, \quad \cdots$$

which corresponds to the even decomposition.

1.5   Special Forms of the Discrete Transform

The discrete transform has some special properties quite distinct from the continuous transform that are of interest to consider. Since it deals with sampled signals, it is inherently associated with a numerical scheme of evaluation. This section lays out some of those properties that are significant for the development of numerical algorithms.

Separate Real and Imaginary Functions

It was seen before that if the function $f(t_m)$ is real and D_n is its transform, then, as a direct consequence of equation (1.8), it can be established that

$$\text{Real } \{D_n\} = \tfrac{1}{2}[D(n) + D(N - n)]$$
$$\text{Imag } \{D_n\} = -\tfrac{1}{2}[D(n) - D(N - n)]$$

That is, the real part is even and the imaginary part is odd. Similarly, if the function $f(t_m)$ is imaginary only, then

$$\text{Real } \{D_n\} = \tfrac{1}{2}[D(n) - D(N - n)]$$
$$\text{Imag } \{D_n\} = \tfrac{1}{2}[D(n) + D(N - n)]$$

What both of these show is that when performing Fourier transforms, information is neither gained nor lost. For example, 1024 pieces of real-only data are converted into 512 pieces of complex data.

Transform of Two Real Functions

To handle the case when $f(t_m)$ is complex, consider it as the sum of two real-only functions as

$$f(m) = g(m) + ih(m)$$

Letting the transforms be represented by upper case letters, then from the linearity property the transform of this is

$$F(n) = G(n) + iH(n)$$
$$= R(n) + iI(n)$$

where $R(n)$, $I(n)$ represent, respectively, the array of real and imaginary elements of the transform $F(n)$. Since the transform of $g(m)$ on its own is even in its real part and odd in its imaginary part, then

$$G(n) = \tfrac{1}{2}[R(n) + R(N - n)] + i\tfrac{1}{2}[I(n) - I(N - n)] \qquad (1.24)$$

Similarly, since the transform of the function $h(m)$ is odd in its real part and even in its imaginary part, then

$$H(n) = \tfrac{1}{2}[I(n) + I(N - n)] - i\tfrac{1}{2}[R(n) - R(N - n)] \qquad (1.25)$$

A useful application of these results is in the efficient computation of the transform of real-only functions. The transform of two real functions can be obtained simultaneously by first forming a complex function from both of them as above. Thus if the real and imaginary parts of the discrete transform of the complex function are decomposed as above, then the simultaneous transform of two real functions can be accomplished. This effectively gives a doubling of the efficiency of the algorithm.

Transform of $2N$ Samples with an N Sample Transform

There may be a limit on the maximum size of the data sample that can be transformed at any one time. For example, the in-core storage may restrict data arrays
to a size of 4096, say. When the data size is larger than this, it is possible, nonetheless, to be able to perform the transform on the larger data size. As will be seen, it
only requires some pre- and post-sorting of the arrays.

Consider a function $f(t_m)$ that is described by $2N$ samples. It is desired to break
it into two separate collections so that only an N transform need be performed. The
function cannot simply be divided in half, instead divide it as follows

$$g(m) = f(2m), \qquad m = 0, 1, 2, \ldots, N - 1$$
$$h(m) = f(2m + 1)$$

That is, it is broken into its even- and odd-numbered parts. (These, of course, are
not the even and odd decomposition of the functions.) The transform can now be
written as

$$D(n) = \Delta T \sum_{m=0}^{2N-1} f(m) e^{-i2\pi nm/2N}$$
$$= \Delta T \sum_{m=0}^{N-1} f(2m) e^{-i2\pi n(2m)/2N} + \Delta T \sum_{m=0}^{N-1} f(2m+1) e^{-i2\pi n(2m+1)/2N}$$
$$= \Delta T \sum g(m) e^{-i2\pi nm/N} + e^{-i\pi n/N} \Delta T \sum h(m) e^{-i2\pi nm/N}$$

It is seen that the summation terms are actually N - point transforms and so can
be written as

$$D(n) = G(n) + e^{-i\pi n/N} H(n), \qquad n = 0, 1, \ldots, 2N - 1 \qquad (1.26)$$

This is a completely general result and shows that if only an N -algorithm is available, nonetheless, transforms of multiple N can be performed. More generally, if
the data size is $8N$ say, then the data can be broken down as 2 sets of $4N$, then 4
sets of $2N$, and finally, 8 sets of N. The N- point algorithm can then be used on
each set and some post-sorting of the arrays is all that is required to reconstruct
the $8N$ transform. This idea is at the heart of the FFT algorithm introduced in the
next section. Notice that it works best when the data size is some power of 2 times
N.

Real-Only Data

One practical use of the above result is in taking the transform of a real-only
function. To accomplish this efficiently, construct a complex function given by

$$f(m) = g(m) + ih(m), \qquad m = 0, 1, \ldots, N - 1$$

where $g(m)$ and $h(m)$ are obtained from the real-only values as the even- and odd-numbered values, respectively. Using the previous form for extracting the transform of two real functions, equations (1.24) & (1.25), now gives

$$
\begin{aligned}
F(n) &= \tfrac{1}{2}[R(n) + R(N - n)] + i\tfrac{1}{2}[I(n) - I(N - n)] \\
&\quad + e^{-i\pi n/N}\{\tfrac{1}{2}[I(n) + I(N - n)] - i\tfrac{1}{2}[R(n) - R(N - n)]\}
\end{aligned}
$$

Thus a $2N$ real point transform is accomplished by an N complex point transform. Of course, there is the need to do some sorting of the results, but this is not very time consuming.

1.6   Fast Fourier Transform Algorithm

The final step in the numerical implementation is the development of an efficient algorithm for performing the summations of the discrete Fourier transform. The fast Fourier transform (FFT) is simply a very efficient scheme for computing the discrete Fourier transform. It is not a different transform. To put it more directly, the numbers obtained from the FFT are exactly the same in every respect as those obtained from the DFT. It is not the intention of this section to describe the algorithmic aspects in detail, but just to survey the major features and give a taste as to how the great speed is achieved. More detailed accounts can be found in the References [18,30,36].

Consider the generic forward transform written as

$$
S_n = \sum_{m=0}^{N-1} f_m e^{-i2\pi \frac{nm}{N}}, \qquad n = 0, 1, \ldots, N - 1
$$

In expanded form, this is written out explicitly as

$$
\begin{aligned}
S_0 &= \{f_0 + f_1 + f_2 + \cdots\} \\
S_1 &= \{f_0 + f_1 e^{-i2\pi\frac{1}{N}} + f_2 e^{-i2\pi\frac{2}{N}} + \cdots\} \\
S_2 &= \{f_0 + f_1 e^{-i2\pi\frac{2}{N}} + f_2 e^{-i2\pi\frac{4}{N}} + \cdots\} \\
&\vdots \\
S_n &= \{f_0 + f_1 e^{-i2\pi\frac{n}{N}} + f_2 e^{-i2\pi\frac{n2}{N}} + \cdots\}
\end{aligned}
$$

and so on. For each S_n there are $(N - 1)$ complex products and $(N - 1)$ complex sums. Consequently, the total number of computations (in round terms) is on the order of $2N^2$. There are also $(N - 1)^2$ complex trigonometric function evaluations, but this can be reduced to $(N - 1)$ by rewriting the sums in the form

$$
S_n = f_0 + W^n[f_1 + W^n[f_2 + W^n[f_3 + \cdots]]]
$$

It is worth noting that while the above requires N^2 computations, this is a big improvement on finding the coefficients by solving simultaneous equations. In that case, the computations are on the order of N^3.

The purpose of the FFT is to take advantage of the special form of the exponential terms to reduce the number of computations to even less than N^2.

Repeated Patterns

The key to understanding the FFT algorithm lies in seeing the repeated forms of numbers. This will be motivated by considering the special cases of N being 4 and 8.

First consider the matrix of the exponents $-i2\pi\left(\frac{mn}{N}\right)$

$$\frac{-i2\pi}{N}\begin{bmatrix} 0 & 0 & 0 & 0 & \cdots & 0 \\ 0 & 1 & 2 & 3 & \cdots & (N-1) \\ 0 & 2 & 4 & 6 & \cdots & 2(N-1) \\ 0 & 3 & 6 & 9 & \cdots & 3(N-1) \\ \vdots & \vdots & \vdots & \vdots & \vdots & \vdots \\ 0 & (N-1) & 2(N-1) & 3(N-1) & \cdots & (N-1)(N-1) \end{bmatrix}$$

It is apparent that for an arbitrary value of N, 2π will be multiplied by non-integer numbers (in general). These exponents, however, can be made quite regular if N is highly composite. For example, if N is one of the following

$$N = 2^7 = 2, 4, 8, 16, 32, 64, 112, 256, 512, 1024, \ldots$$

then the effective number of different integers in the matrix is decreased. Thus if $N = 4$

$$\frac{-i2\pi}{4}\begin{bmatrix} 0 & 0 & 0 & 0 \\ 0 & 1 & 2 & 3 \\ 0 & 2 & 4 & 4+2 \\ 0 & 3 & 4+2 & 4+4+1 \end{bmatrix} = \frac{-i2\pi}{4}\begin{bmatrix} 0 & 0 & 0 & 0 \\ 0 & 1 & 2 & 3 \\ 0 & 2 & 0 & 2 \\ 0 & 3 & 2 & 1 \end{bmatrix}$$

This comes about because the exponentials take on the following simple forms

$$e^{-i2\pi[0]} = e^{-i2\pi[1]} = e^{-i2\pi[2]} = e^{-i2\pi[3]} = 1$$

The regularity in the numbers is even more striking for the case of $N = 8$, which has

$$\frac{-i2\pi}{8}\begin{bmatrix} 0 & 0 & 0 & 0 & 0 & 0 & 0 & 0 \\ 0 & 1 & 2 & 3 & 4 & 5 & 6 & 7 \\ 0 & 2 & 4 & 6 & 0 & 2 & 4 & 6 \\ 0 & 3 & 6 & 1 & 4 & 7 & 2 & 5 \\ 0 & 4 & 0 & 4 & 0 & 4 & 0 & 0 \\ 0 & 5 & 2 & 7 & 4 & 1 & 6 & 3 \\ 0 & 6 & 4 & 2 & 0 & 6 & 4 & 2 \\ 0 & 7 & 6 & 5 & 4 & 3 & 2 & 1 \end{bmatrix}$$

It is seen that many of the computations used in forming one of the summations is also used in the others. For example, S_0, S_2, S_4, S_6 all use the sum $(f_0 + f_4)$. The regularity is enhanced even more if 4 $(N/2)$ is added to the latter part of the odd rows. That is, if it is written as

$$\frac{-i2\pi}{8} \begin{bmatrix} 0 & 0 & 0 & 0 & (0 & 0 & 0 & 0) + 0 \\ 0 & 1 & 2 & 3 & (0 & 1 & 2 & 3) + 4 \\ 0 & 2 & 4 & 6 & (0 & 2 & 4 & 6) + 0 \\ 0 & 3 & 6 & 1 & (0 & 3 & 6 & 1) + 4 \\ 0 & 4 & 0 & 4 & (0 & 4 & 0 & 4) + 0 \\ 0 & 5 & 2 & 7 & (0 & 5 & 2 & 7) + 4 \\ 0 & 6 & 4 & 2 & (0 & 6 & 4 & 2) + 0 \\ 0 & 7 & 6 & 5 & (0 & 7 & 6 & 5) + 4 \end{bmatrix}$$

Realizing that $e^{-i2\pi 4/8} = -1$, then it is seen that all the odd summations contain common terms such as $(f_0 - f_4)$. This reuse of the same computations is the first great reduction afforded by the FFT.

Binary Representation

The trick to understanding the pattern of repeated numbers in the FFT algorithm lies is the binary representation of these numbers. In this number scheme a general integer is pieced together from the collection 1, 2, 4, 8, 16, 32, ... For example, the numbers 7, 11 can be written as

$$7 = 4 + 2 + 1, \qquad 11 = 8 + 2 + 1$$

By introducing a set of binary coefficients (i.e., they can only have values 0 or 1), then the above integers can be written as

$$m = m_8 8 + m_4 4 + m_2 2 + m_1 1, \qquad m_8, m_4, m_2, m_1 = 0, 1$$

Thus 7 has the four coefficients 0111, while 11 has the four 1011. Of course, as larger numbers need to be represented, then more coefficients need to be added. Each extra coefficient adds another level, thus 16 has 4 levels, while 1024 has 10. The level of a number is obtained from

$$\text{level:} \qquad \gamma = \log_2(n)$$

The summations will now be replaced with their binary representations. For convenience, introduce the notation for what are called the *twiddle factors*

$$e^{-i2\pi \frac{nm}{N}} = W^{nm}, \qquad W \equiv e^{-i2\pi/N}$$

and replace the subscripts as indexes, then

$$S(n) = \sum_{m=0}^{N-1} f(m)W^{mn}$$

Replace all integers by their binary representation also, giving (for concreteness, the following development will be done for $N = 8$)

$$n = n_4 4 + n_2 2 + n_1 1, \qquad m = m_4 4 + m_2 2 + m_1 1$$

Then the exponent of the twiddle factor becomes

$$\begin{aligned}
nm &= (n_4 4 + n_2 2 + n_1 1)(m_4 4 + m_2 2 + m_1 1) \\
&= (n_4 16 + n_2 8 + n_1 4)m_4 + (n_4 8 + n_2 4 + n_1 2)m_2 + (n_4 4 + n_2 2 + n_1 1)m_1
\end{aligned}$$

Since exponents of multiple N gives $W^{16} = W^8 = 1$, then the general form reduces to

$$W^{nm} = W^{(n_1 1)m_4 4} W^{(n_2 2 + n_1 1)m_2 2} W^{(n_4 4 + n_2 2 + n_1 1)m_1 1}$$

The summation over eight terms can now be broken into three binary summations as

$$S(n_4 4 + n_2 2 + n_1 1) =$$

$$\sum_{m1=0}^{1} \left[\sum_{m2=0}^{1} \left\{ \sum_{m4=0}^{1} f(m_4 4 + m_2 2 + m_1 1)W^{(n_1 1)m_4 4} \right\} W^{(n_2 2 + n_1 1)m_2 2} \right] W^{(n_4 4 + n_2 2 + n_1 1)m_1}$$

In general, the sum over N terms is replaced with $2 \times level$ sums. Now look more specifically at the partial sums indicated by the brackets and braces:

$$S_1(n_1 4 + m_2 2 + m_1 1) = \sum_{m4=0}^{1} f(m_4 4 + m_2 2 + m_1 1)W^{(0)} W^{n_1 m_4 4}$$

$$S_2(n_1 4 + n_2 2 + m_1 1) = \sum_{m2=0}^{1} S_1(n_1 4 + m_2 2 + m_1 1)W^{(n_1 1)m_2 2} W^{n_2 m_2 4}$$

$$S_3(n_4 4 + n_2 2 + n_1 1) = \sum_{m1=0}^{1} S_2(n_1 4 + n_2 2 + m_1 1)W^{(n_2 2 + n_1 1)m_1} W^{n_4 m_4 4}$$

Noting that $W^{k4} = \pm 1$ depending on k being 0 or 1, respectively, then the above can be further expanded to

$$\begin{aligned}
S_1(0 + m_2 2 + m_1 1) &= f(0 + m_2 2 + m_1 1) + f(4 + m_2 2 + m_1 1)W^{(0)} \\
S_1(4 + m_2 2 + m_1 1) &= f(0 + m_2 2 + m_1 1) - f(4 + m_2 2 + m_1 1)W^{(0)}
\end{aligned}$$

$$\begin{aligned}
S_2(n_1 4 + 0 + m_1 1) &= S_1(n_1 4 + 0 + m_1 1) + S_1(n_1 4 + 2 + m_1 1)W^{(n_1 2)} \\
S_2(n_1 4 + 2 + m_1 1) &= S_1(n_1 4 + 0 + m_1 1) - S_1(n_1 4 + 2 + m_1 1)W^{(n_1 2)}
\end{aligned}$$

$$\begin{aligned}
S_3(n_1 4 + n_2 2 + 0) &= S_2(n_1 4 + n_2 2 + 0) + S_2(n_1 4 + n_2 2 + 1)W^{(n_2 2 + n_1 1)} \\
S_3(n_1 4 + n_2 2 + 1) &= S_2(n_1 4 + n_2 2 + 0) - S_2(n_1 4 + n_2 2 + 1)W^{(n_2 2 + n_1 1)}
\end{aligned}$$

These equations form the heart of the FFT algorithm. Notice how the partial sums at each level feeds into the next. At each level, the binary coefficients are treated as dummy variables rangeing from 0 to 1 allowing the above to be interpreted as a system of equations.

There are $\log_2(N)$ levels, and in each there are $(N-1)$ complex sums and $N/2$ complex products. Consequently, the number of computational loops with and without the FFT algorithm is given by

$$\frac{3}{2}N \log_2 N \qquad \text{versus} \qquad 2N^2 \qquad\qquad (1.27)$$

When $N = 8$, this gives a speed factor of only 3.5:1, but when $N = 1024$, this jumps to over 100:1, and when $N = 4096$ it gives a whopping 400:1 performance. It is this excellent performance at large N that makes the application of Fourier analysis feasible for practical problems. Note that the indexes are grouped in pairs. For example, in level 2

$$S_2(0) = S_1(0) + S_1(2)W^0$$
$$S_2(2) = S_1(0) + S_1(2)W^0$$

In fact at each level the indexes of the pairs differ by $N/2^{level}$. What this means is that during the computations, the new values can be interchanged with the old. This in-place exchange can be efficient on storage.

Unscrambling the Transform

The final step in the process is to unscramble the result. The algorithm provides the array $S_3(\ )$, but this is related to the desired values by

$$S(n_4 4 + n_2 2 + n_1 1) = S_3(n_1 4 + n_2 2 + n_4 1) \qquad\qquad (1.28)$$

The unscrambling, therefore, is easily accomplished by just running through the binary coefficients . That is, ranging over n_4, n_2, n_1 gives

$$(n_4 4 + n_2 2 + n_1 1) = 0, 1, 2, 3, 4, 5, 6, 7$$
$$(n_1 4 + n_2 2 + n_4 1) = 0, 4, 2, 6, 1, 5, 3, 7$$

These numbers are associated vertically, thus $S(4) = S_3(1)$ and so on.

Computer Program

There are many codes available for performing the FFT as can be found in Reference [191] and well-documented FORTRAN routines are described in Reference [184]. The computer program CFFTCOMP used in all the examples to follow has its source code listed in Appendix D. While the core of the program is the FFT GROUP, it is necessary to have additional features that facilitate the input and output of the data. The basic structure of the program is that three-column data (time, real amplitude, imaginary amplitude) are read in. These data need not be equispaced since the program will interpolate as needed. The number of transform points and ΔT are user chosen. There is then a choice as to how to store the results. The most useful form (and the one used in the programs of the later chapters) is the three-column form of frequency, real part and imaginary part. The inverse transform is made as symmetric as possible with the forward. Again, three-column data are input and the result is stored also as three columns.

To enhance efficiencies, switches in the program allow handling data that are real-only or are stored without their complex conjugates. The program is written in menu style, but obviously the most efficient way of using it is through driver files, i.e., a file containing all the responses needed. Creating these driver files is a built in feature of the program.

The program is written with the personal computer in mind. An example of its performance on an 8 MHz IBM®-AT compatible machine (using an 80287 math co-processor) gives a 4096 complex transform in 12.5 seconds. The corresponding time for a 1024 real transform is 2.1 seconds. These times are given to emphasize that the process of transforming, both forward and inverse, is not a significant time consumer.

1.7 Examples Using the FFT Algorithm

The following examples serve to show the basic procedures used in applying the FFT to transient signals. As will be seen, the major effort is expended in interpreting the results. That is, the extent to which the DFT only approximates the corresponding continuous transform must always be kept in mind.

Rectangular Pulse

Consider the rectangular pulse already treated and that has the transform given by equation (1.4). The effects of different sample lengths and number of points used are shown in Figure 1.4. Since the number of points used is relatively small, then it is necessary to treat the jump correctly. This is done by using its half value as shown in the inset.

First it is noticed that the transform is symmetric about the middle or Nyquist frequency. This is a consequence of the input signal being real-only — if it was complex, then the transform would fill the complete range. What this means is that N real points are transformed into $N/2$ complex points and no information is gained or lost. Therefore, the useful frequency range extends only up to the

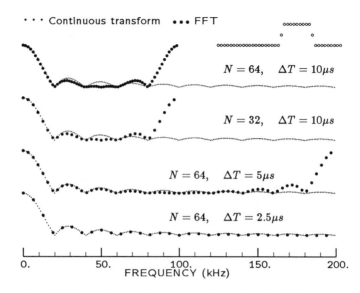

Figure 1.4: FFT transform of a rectangular pulse of width $a = 50\mu s$.

Nyquist, given by

$$f_{\text{Nyquist}} = \frac{1}{2\Delta T}$$

This range is increased only by decreasing ΔT. Thus for a fixed number of points, fine resolution in the time domain (small ΔT) means course resolution in the frequency domain. Finer resolution in the frequency domain is achieved only by increasing the sample length.

It is also seen from the figure that the match between the FFT amplitude and the continuous Fourier transform is very good at low frequencies but gets worse at the higher frequencies. As mentioned before, the discrete and the continuous transforms match exactly only if the highest significant frequency in the signal is less than the Nyquist. The rectangular pulse can be represented exactly only by using an infinite number of sinusoids but the contributing amplitudes get smaller as frequency increases.

For a given sample rate ΔT, the number of samples only determine the density of transform points. Thus the top two transforms in Figure 1.4 are numerically identical. Increasing the sample rate for a given number of samples increases the Nyquist frequency, and therefore the range of comparison between the discrete and continuous transforms. It is seen that by the fourth plot, the amplitudes in the vicinity of the Nyquist are negligible. Using the discrete transform puts an upper limit on the maximum frequency available to characterize the signal. If the signal is not smooth, the amplitudes of the high-frequency sinusoids used to describe the signal are high. Any attempt at capping the high-frequency sinusoids introduces a distortion in the amplitudes of the lower-frequency sinusoids. This is called *aliasing* and the next section discusses it in more detail.

Modulated Function

Consider the cosine wave modulated by the rectangular pulse in the form

$$f(t) = \cos(qt) \qquad -t_o < t < t_o + a$$
$$= 0 \qquad \text{otherwise}$$

This is shown as an inset in Figure 1.5. By adjusting q, the number of zeros can be increased or decreased within the time band a. The transform of the modulated pulse is also shown in Figure 1.6. The Nyquist frequency is at 100 kHz.

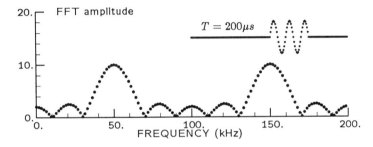

Figure 1.5: Transform of a modulated pulse with $q = 50$ kHz and $a = 50\mu s$.

It is apparent what is happening. The cosine function on its own would give a spike at $\omega = q$ (50 kHz in this example), whereas the rectangle on its own is fairly broad with side lobes (as in Figure 1.4). The product of the two in the time domain is a convolution in the frequency domain. The convolution tends to distribute the effect of the pulse. A point of interest here is that if a narrow-banded signal in the frequency domain is desired, then it is necessary to make the time domain signal of the form of this example (i.e., having many zeros.) Otherwise, it will appear broad-banded.

1.8 Sampled Waveforms

By necessity, real-signal records are finite in length, and this can add other problems to the scheme of estimating its spectral content. This section reviews the major effects that should be considered.

Aliasing

The basic difference between the continuous and the discrete Fourier transforms is sampling (both rate and duration). An actual signal of very long duration must be truncated to a reasonable size before the discrete Fourier transform is evaluated. The transform of the signal will therefore be dependent on the actual function $f(t)$ and the way it is sampled. Note that if the sampling rate is every ΔT, then the highest detectable frequency using the FFT is $1/(2\Delta T)$. Thus, if ΔT is too large, the high frequency appears as a lower frequency (its alias) as shown in Figure 1.6. This phenomenon of aliasing can be avoided if the sampling rate is high enough. By 'high enough' is implied a rate commensurate with the highest significant frequency in the signal.

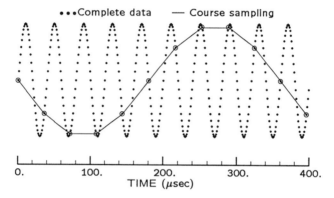

Figure 1.6: Causing aliasing by course sampling.

From a practical point of view, this is also avoided by recording with the analog filters set so as to remove significant frequencies above the Nyquist.

Leakage

The estimated spectrum is given only at discrete frequencies and this can cause another problem known as leakage. For example, suppose there is a spectral peak at 17 kHz and the sampling is such that the spectrum values are only at every 5 kHz. Then the energy associated with the 17 kHz peak leaks into the neighboring frequencies, thus distorting their spectral estimates.

Leakage is most significant in instances where the signal has sharp spectral peaks and is least significant in cases where the signal is a flat, broad-banded signal. In the analysis of the impact of structures, the signals generated generally do not exhibit spectral peaks (as in the case of vibrations). However, it can become a problem if there are many reflections present in the signal.

Truncation and Windowing

When the signal to be analyzed is completely contained in the sampling time window (which means it is periodic), there is no leakage; but when a window (such as a rectangular window) is used to analyze a signal of duration more than $T = (N - 1)\Delta T$, leakage will be present. The effect of leakage can be reduced if the window makes the signal look "more periodic" in the time record T. This is done by using a window that will smoothly attenuate the signal at the ends of the sampled time record.

The choice of a window is dictated by the type errors induced by using the window (e.g., amplitude error or frequency error). It is to be remembered that no window will eliminate leakage completely while introducing no errors in the amplitude or frequency resolution. Examples of windows commonly used are the Hanning window and the rectangular window (see References [21,33] for more details). The window used in most of the studies to follow is the rectangular window tapered in some way on the large time side. This window is thought to be adequate and is certainly the easiest to implement.

Padding

Since discrete Fourier analysis represents a finite sample of an infinite signal on a finite period, then schemes for increasing the apparent period must be used. The simplest means of doing this is by padding. There are various ways of padding the signal and the one most commonly used is that of simply adding zeroes. Alternatively, Reference [57] shows how a long term estimate of the signal behavior can also be used to pad the signal. The main consequence of padding is that since a larger time window is used then higher resolution in the frequency domain is achieved. It also makes the signal non-periodic over a larger time domain. These effects can be seen in Figure 1.4.

As regards padding with zeros it should be pointed out that it really doesn't matter if the zeros are before the recorded signal or after. Thus even if the signal never decreases to zero, padding is justified because it represents the early quiescent part of the signal.

Chapter 2

Spectral Analysis of Wave Motion

The application of the continuous Fourier transform to the solution of dynamic problems is quite standard as can be judged from some of the chapters of Reference [205]. The application of Fourier series, however, is quite limited with the exceptions being References [37,49,104,200]. It appears that the task of performing the summation is too formidable since a large number of terms are required to adequately describe the wave. With the advent of computers and the utilization of the FFT algorithm, the picture changed, and performing large summations rapidly is now feasible and economical.

This chapter introduces, in a consistent manner, the application of the discrete Fourier transform (in the form of spectral analysis) to the solution of wave problems. The crucial step is to set up the connection between the spectral responses at different space locations. In doing so, certain key ideas emerge, central among which are those of the spectrum relation and the group behavior.

This chapter also summarizes some of the experimental aspects of waves in structures and how the time series measurements can be used for spectral estimation. These aspects are introduced in this early chapter as a reminder that the spectral approach finds its most fruitful application in the analysis of experimental data from real systems. Some practical considerations of using the FFT for the inversion process are also given.

2.1 Spectral Analysis of Differential Equations

The key to the spectral description of waves is to be able to express the phase changes incurred as the wave propagates from location to location. This is done conveniently through use of the governing differential equations for particular structural models (although other schemes are possible). It is done in this chapter, however, in a way that considers the general properties of the solutions independent of particular structural applications. (Specific applications are left to the remaining chapters.)

The idea of representing the time variation of a function by a summation of harmonic functions is extended here to representing arbitrary functions of time and position resulting from the solution of wave equations. The approach is to remove the time variation by using the spectral representation of the solution. This leaves a new differential equation for the coefficients, which in many cases can be integrated directly.

General Functions of Space and Time

The solutions in wave problems are general functions of space and time. If the time variation of the solution is focused on at a particular point in space, then it has the spectral representation

$$u(x_1, y_1, t) = f_1(t) = \sum C_{1n} e^{i\omega_n t}$$

At another point, it behaves as a time function $f_2(t)$ and is represented by the Fourier coefficients C_{2n}. That is, the coefficients are different at each spatial point. Thus, the solution at an arbitrary position has the following spectral representation

$$u(x, y, t) = \sum \hat{u}_n(x, y, \omega_n) e^{i\omega_n t} \tag{2.1}$$

where $\hat{u}_n$ are the spatially dependent Fourier coefficients. Notice that these coefficients are functions of frequency ω and thus there is no reduction in the total number of independent variables.

For shorthand, the summation and subscripts will often be understood and the function will be given the representation

$$u(x, y, t) \Rightarrow \hat{u}_n(x, y, \omega_n) \qquad \text{or} \qquad \hat{u}(x, y, \omega)$$

Sometimes it will be written simply as $\hat{u}$.

Derivatives

The differential equations are given in terms of both space and time derivatives. Since they are linear, then it is possible to apply the spectral representation to each term appearing. Thus, the spectral representation for the time derivative is

$$\frac{\partial u}{\partial t} = \frac{\partial}{\partial t} \sum \hat{u}_n e^{i\omega_n t} = \sum i\omega_n \hat{u}_n e^{i\omega_n t}$$

In shorthand, this becomes

$$\frac{\partial u}{\partial t} \Rightarrow i\omega_n \hat{u}_n \qquad \text{or} \qquad i\omega \hat{u}$$

In fact, time derivatives of general order have the representation

$$\frac{\partial^m u}{\partial t^m} \Rightarrow i^m \omega_n^m \, \hat{u}_n \qquad \text{or} \qquad i^m \omega^m \, \hat{u} \qquad (2.2)$$

Herein lies the advantage of the spectral approach to solving differential equations — algebraic expressions in the Fourier coefficients replace the time derivatives. That is, there is a reduction in the number of derivatives occurring.

Similarly, the spatial derivatives are represented by

$$\frac{\partial u}{\partial x} = \frac{\partial}{\partial x} \sum \hat{u}_n e^{i\omega_n t} = \sum \frac{\partial \hat{u}_n}{\partial x} e^{i\omega_n t}$$

and in shorthand notation

$$\frac{\partial u}{\partial x} \Rightarrow \frac{\partial \hat{u}_n}{\partial x} \qquad \text{or} \qquad \frac{\partial \hat{u}}{\partial x} \qquad (2.3)$$

In this case there appears to be no reduction, but as will be seen later, with the removal of time as an independent variable, these derivatives often become ordinary derivatives, and thus more amenable to integration.

Differential Equations

Consider the following general, linear (one-dimensional), homogeneous differential equation in the dependent variable $u(x,t)$

$$u + a\frac{\partial u}{\partial x} + b\frac{\partial u}{\partial t} + c\frac{\partial^2 u}{\partial x^2} + d\frac{\partial^2 u}{\partial t^2} + e\frac{\partial^2 u}{\partial x \partial t} + \ldots = 0 \qquad (2.4)$$

The coefficients $a, b, c\ldots$ are assumed not to depend on time, but could be functions of position. If, now, the solution is given the spectral representation

$$u(x,t) = \sum_n \hat{u}_n(x, \omega_n) e^{i\omega_n t}$$

then on substitution into the differential equation get

$$\sum_n \left\{ \hat{u}_n + a\frac{d\hat{u}_n}{dx} + (i\omega_n)b\hat{u}_n + c\frac{d^2\hat{u}_n}{dx^2} + (i\omega)^2 d\hat{u}_n + (i\omega)e\frac{d\hat{u}_n}{dx} + \ldots \right\} e^{i\omega_n t} = 0$$

Since each $e^{i\omega_n t}$ term is independent, then this equation must be satisfied for each n. That is, there are n simultaneous equations of the form

$$\left[1 + (i\omega_n)b + (i\omega_n)^2 d + \ldots\right] \hat{u}_n + \left[a + (i\omega_n)e + \ldots\right] \frac{d\hat{u}_n}{dx} + \left[c + \ldots\right] \frac{d^2\hat{u}_n}{dx^2} + \ldots = 0$$

These equations become, on grouping terms,

$$A_1(x, \omega)\hat{u} + A_2(x, \omega)\frac{d\hat{u}}{dx} + A_3(x, \omega)\frac{d^2\hat{u}}{dx^2} + \ldots = 0 \qquad (2.5)$$

where $A_1, A_2, \ldots$ depend on frequency and position and are complex. The subscript n is dropped, but it will be understood that an equation of the above form must be solved at each frequency. It is seen that the original partial differential equation becomes a set of ordinary linear differential equations in the Fourier coefficients $\hat{u}$.

Spectrum Relation

Consider the special case when the coefficients $a, b, c, \ldots$ in the partial differential equation are independent of position. (Variable coefficients will be treated in the later chapters.) As an aside, however, it should be realized that in practical problems there is always a spatial region over which the coefficients are sensibly constant. That is, they can be replaced by their average constant value and so the results to follow are applicable and give some intimation of their behaviors.

Linear differential equations with constant coefficients have solutions of the form $e^{\lambda x}$, where λ is obtained by solving the algebraic characteristic equation

$$A_1 + A_2\lambda + A_3\lambda^2 + \ldots = 0 \tag{2.6}$$

Reference [108] considers in good detail the solution of equations of this form and should be consulted for the details not given here. It is usual in wave analysis, however, to assume that λ is complex to begin with, that is, that the solutions are of the form

$$\hat{u}(x) = C e^{ikx} \tag{2.7}$$

For example, consider the differential equation

$$a\hat{u} + \frac{d\hat{u}}{dx} = 0 \quad \Rightarrow \quad [a + ik]C = 0$$

This gives k and the solution as

$$k = -a$$
$$\hat{u}(x) = C e^{-ax}$$

where C is a constant of integration. Similarly, the following second-order differential equation

$$au + \frac{d^2u}{dx^2} = 0 \quad \Rightarrow \quad [a - k^2]C = 0$$

gives k and the solution as

$$k = \pm\sqrt{a}$$
$$\hat{u}(x) = C_1 e^{i\sqrt{a}x} + C_2 e^{-i\sqrt{a}x}$$

There are two solutions (and constants of integration) because k occurred to the power of two. Note that, even if the coefficients in the differential equation are real, it is possible for k to be complex.

In general, then, the characteristic equation becomes

$$A_1 + (ik)A_2 + (ik)^2 A_3 + \ldots = 0$$

and this has many values of k that satisfies it. That is,

$$k_{mn} = f_m(A_1, A_2, A_3 \ldots, \omega_n) \tag{2.8}$$

This relation between the exponent k (called the *wavenumber*) and frequency ω is called the *spectrum relation* and is fundamental to the spectral analysis of waves. The different values of m correspond to the different modes. The solution is given as the superposition of modes in the form

$$\hat{u}(x) = C_1 e^{ik_1 x} + C_2 e^{ik_2 x} + \ldots + C_m e^{ik_m x}$$

There are as many modes (or solutions) are there are roots of the characteristic equation and these should not be confused with the number of solutions at each frequency. To reinforce this, the solution in total form is written as

$$u(x,t) = \sum_n \left\{ C_{1n} e^{ik_{1n} x} + C_{2n} e^{ik_{2n} x} + \ldots C_{mn} e^{ik_{mn} x} \right\} e^{i\omega_n t} \tag{2.9}$$

The exponential form for each term is due to the coefficients of the differential equation being constant; however, the solution for any problem can always be expressed as

$$u(x,t) = \sum F_n G(k_{mn} x) e^{i\omega_n t} \tag{2.10}$$

where F_n is an amplitude spectrum and G (which may be a combination of modes) is the system transfer function. Analysis of the partial differential equations combined with the boundary conditions determine the particular forms for G and $k_m(\omega)$. In fact, in comparison to the results of Chapter 1, it is seen that G determines the phase shifts with respect to space. Further, it is noted that the wavenumber k acts as a scale factor on the position variable in the same way that the frequency ω acts on the time. The analysis of the scaling done by $k(\omega)$ provides a good deal of insight into the solution before the actual solution is obtained.

2.2   Examples

The following are some simple examples to demonstrate the effectiveness of the spectral approach to dynamic equations. Keep in mind that the net result in each case is to replace the original set of partial differential equations by a collection of undetermined coefficients. These coefficients are then determined, subsequently, by imposing the appropriate initial and boundary conditions.

Simple Wave Equation

Consider the following differential equation which is an example of the simple 1-D wave equation

$$\frac{\partial^2 u}{\partial x^2} = a^2 \frac{\partial^2 u}{\partial t^2} \tag{2.11}$$

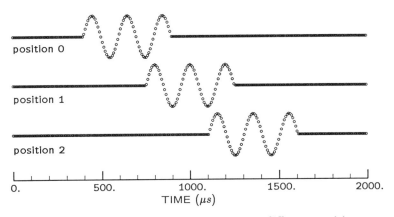

Figure 2.1: Segments of an infinite wavetrain at different positions

where a is real and positive. Using the spectral representation gives

$$\frac{d^2\hat{u}}{dx^2} + \omega^2 a^2 \hat{u} = 0$$

Since this has constant coefficients, assume Ce^{ikx} is a solution which (on substitution) results in the characteristic equation

$$\left[-k^2 + \omega^2 a^2\right] C = 0$$

Consequently, the two modes for k are

$$k_1(\omega) = +\omega a, \qquad\qquad k_2(\omega) = -\omega a \qquad\qquad (2.12)$$

and the solution is

$$u(x,t) = \sum\{C_1 e^{i\omega ax} + C_2 e^{-i\omega ax}\}e^{i\omega t} = \sum C_1 e^{i\omega(ax+t)} + \sum C_2 e^{-i\omega(ax-t)}$$

These correspond to sinusoids (infinite wave trains) moving to the left and right, respectively. A line in the $x - t$ plane connecting a common point is a straight line. That is, points of common phase (position on the sinusoid) travel along straight lines in $x - t$ as shown in Figure 2.1 for a forward moving train. If the slope of this line is thought of as a speed, then it can be called the phase speed and is given by

$$c = \frac{x}{t} = \frac{1}{a} = \frac{\omega}{k}$$

It is seen that different frequency components travel at the same speed. When the phase speed is constant with respect to frequency, the signal is said to be *non-dispersive* and it will maintain its superposed shape. That is, at each different position in space, the sinusoids have the same relative phase shift and so will superpose to give the same shape.

The relation between phase speed and frequency is called the *dispersion relation*. The set of coefficients $C_1(\omega), C_2(\omega)$ are called the *amplitude spectrums*.

Simple Dispersive Equation

Consider the following differential equation (which actually represents flexural wave motion in beams)

$$\frac{\partial^4 u}{\partial x^4} + a^4 \frac{\partial^2 u}{\partial t^2} = 0 \qquad (2.13)$$

where, again, a is real and positive. Since the coefficients are constant, assume $Ce^{-i(kx-\omega t)}$ is a solution giving the characteristic equation as

$$[k^4 - \omega^2 a^4]\, C = 0$$

Now, however, that there are four modes since

$$k^2 = \pm \omega a^2$$

$$k_1(\omega) = +a\sqrt{\omega}, \quad k_2(\omega) = -a\sqrt{\omega}, \quad k_3(\omega) = +ia\sqrt{\omega}, \quad k_4(\omega) = -ia\sqrt{\omega} \qquad (2.14)$$

and the total solution becomes

$$u(x,t) = \sum C_1 e^{i(a\sqrt{\omega}x + \omega t)} + \sum C_2 e^{-i(a\sqrt{\omega}x - \omega t)} + \sum C_3 e^{-a\sqrt{\omega}x + i\omega t} + \sum C_4 e^{a\sqrt{\omega}x + i\omega t}$$

The first two are wave terms, as in the previous example, and again travel with a constant phase speed given by

$$c = \frac{\omega}{k} = \frac{1}{a}\sqrt{\omega}$$

Different frequency components, however, travel at different speeds. As a result, the superposition of the components changes at the different locations and the wave will appear to change shape as it propagates. When the phase speed is not constant with respect to frequency, the signal is said to be *dispersive*. This is demonstrated for two wave-train segments shown in Figure 2.2. Observe that while the amplitudes of the individual components remain the same, the amplitude of the resultant signal varies considerable in time.

The last two terms are spatially damped vibrations. While not wave solutions *per se*, they nonetheless play a crucial role in being able to satisfy the imposed conditions when, say, a wave interacts with a boundary.

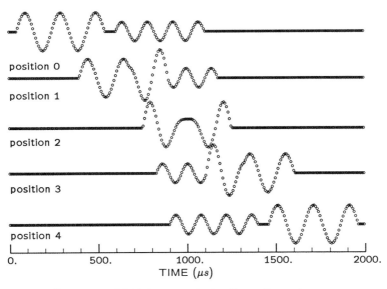

Figure 2.2: Segments of infinite wave trains for a dispersive response.

Thermal Conduction

The two previous examples had the familiar double time derivatives associated with inertia. Consider, as a different case, the following differential equation related to Fourier's law of heat conduction

$$\frac{\partial^2 u}{\partial x^2} = a \frac{\partial u}{\partial t} \tag{2.15}$$

where a is real and positive. Again, since the coefficients are constant, assume $Ce^{-i(kx-\omega t)}$ is a solution giving the characteristic equation as

$$[k^2 - i\omega a] C = 0$$

There are two modes having the spectrum relations

$$k(\omega) = \pm\sqrt{-ia\omega} = \pm(1-i)\sqrt{a\omega/2} \equiv \pm(1-i)\beta \tag{2.16}$$

and the complete solution is

$$u(x,t) = \sum C_1 e^{\beta x} e^{i(\beta x + \omega t)} + \sum C_2 e^{-\beta x} e^{-i(\beta x - \omega t)}$$

This is comprised of left and right propagating terms, but both are also spatially damped. Further, the solutions are dispersive in the sense that β is not a linear function of ω.

Although this solution does not fall neatly into a simple wave description, the spectral approach solves it with ease. This is very fortunate, because as waves interact with boundaries and discontinuities, they lose their "wave nature" causing a more complicated dynamic motion and the spectral approach can still handle these cases.

Variable Coefficients

As a final example, consider the following differential equation governing axisymmetric wave propagation

$$\nabla^2 u = a^2 \frac{\partial^2 u}{\partial t^2}, \qquad \nabla^2 = \frac{\partial^2}{\partial r^2} + \frac{1}{r}\frac{\partial}{\partial r} \qquad (2.17)$$

This is an example of non-constant coefficients. The spectral representation of the time variation gives

$$\left[\frac{d^2}{dr^2} + \frac{1}{r}\frac{d}{dr} + \beta^2\right]\hat{u} = 0, \qquad \beta = a\omega$$

This has solutions in terms of the Bessel functions $J_o(\beta r)$ and $Y_o(\beta r)$ as seen from Appendix B. The intent of this example, however, is to show the character of the solution. So to that end, let the solution near $r = R$ be only of interest, and replace the variable coefficient r in the second term with its average value R and treat the equation as having constant coefficients. The spectrum relation is then

$$[k^2 + \frac{i}{R}k - \beta^2]C = 0 \qquad \text{or} \qquad k(\omega) = \pm\sqrt{(a^2\omega^2 - \frac{1}{4R^2})} - \frac{i}{2R} \qquad (2.18)$$

It is apparent that this is a damped wave propagation (which indeed is the character of the Bessel solution). Of particular interest, however, is the possibility that the radical could change sign if the frequency is low enough. That is, when

$$a^2\omega^2 < \frac{1}{4R^2} \qquad \text{or} \qquad \omega < \frac{1}{2aR}$$

all terms are complex and so the motion is a completely damped vibration. For frequencies above this value, there is also a propagating component. This special frequency is referred to as the *cut-off frequency*.

This example serves to demonstrate that it is possible to have a wide variety of behaviors under the general umbrella of spectral analysis.

2.3   Propagating and Reconstructing Waves

The significance of the spectral approach to waves (coupled with the use of the differential equations) is that once the signal is characterized at one space position then it is known at all positions, and therefore propagating it becomes a fairly simple matter. This section illustrates the basic algorithm for doing this.

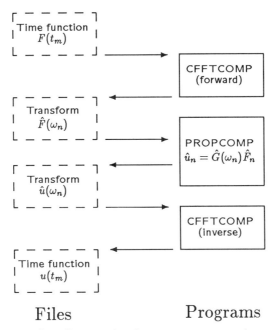

Files   Programs

Figure 2.3: Flow diagram for the wave reconstruction program.

Basic Algorithm

In its simplest terms, the solution to a waves problem is represented as

$$u(x,t) = \sum_n \hat{F}_n \left\{ G_1(k_{1n}x) + G_2(k_{2n}x) + \ldots \right\} e^{i\omega_n t} = \sum \hat{F}_n G(k_{mn}x) e^{i\omega_n t}$$

where G is the analytically known transfer function of the problem. It is a function of position x and has different numerical values at each frequency. $\hat{F}_n$ is the amplitude spectrum; this is known from the input conditions or from some measurement. Thus $F_n G$ is recognized as the Fourier transform of the solution. Of course, it is different at each position, but once it is evaluated at a particular position then its inverse immediately gives the time history of the solution at that point. Figure 2.3 is a flow diagram for its evaluation. Briefly, the time input $F(t)$ is converted to its spectrum $\hat{F}_n$ through a use of the forward FFT. The transformed solution is then obtained by evaluating the product

$$\hat{u}_n = \hat{F}_n G(k_{mn})$$

at each frequency. This is finally reconstructed in the time domain by use of the inverse FFT. In the process, it is necessary to realize (when using the FFT to perform the inversion) that $F_n G$ is evaluated only up to the Nyquist frequency and the remainder is obtained by imposing that it must be the complex conjugate of the initial part. This ensures that the reconstructed time history is real only.

A stripped down version of this basic algorithm coded in FORTRAN is given in program PROPCOMP in Appendix D. (A full fledged example of a propagation program is given in RODCOMP also in Appendix D.) The coding is very simple — although the problem is a time domain convolution, in the frequency domain it is a simple algebraic product. Thus the structure of the program is simply to read in the forcing term, multiply it by the transfer function, and then store it back to disk. An additional feature of note is that when doing propagations the d.c. component is undetermined (since it does not propagate.) It is advantageous to remove its arbitrariness by imposing that the first value be zero.

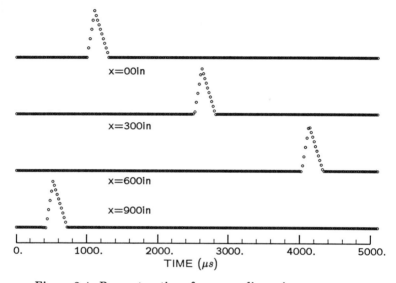

Figure 2.4: Reconstructions for a non-dispersive wave.

Propagating Non-Dispersive Signals

Let the signal history at a particular position be a triangular pulse as shown in Figure 2.4 and let $\hat{u}_n(x_o)$ be its transform. Also let the transfer function be that of the simple wave equation.

$$G_n = e^{-ik_n x}, \qquad k_n = \omega_n/c_o$$

where c_o is a constant. Therefore, the amplitude spectrum is $\hat{F}_n = \hat{u}(x_o)/e^{-ik_n x_o}$ and the solution is then

$$\hat{u}(x) = \hat{u}(x_o)[e^{-ik_n x}]/[e^{-ik_n x_o}]$$

Figure 2.4 shows the reconstructions of the signal at different positions. It is obvious that the signal is propagating at a constant speed and that there is no change in shape. This is characteristic of non-dispersive signals. The amplitude spectrum is shown in Figure 2.5, and it is obvious that the frequency range is adequate to describe the spectral content of the wave.

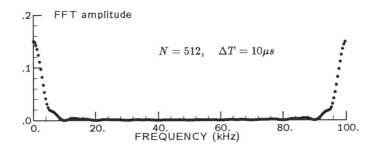

Figure 2.5: Amplitude spectrum for the input signal.

To characterize the amplitude spectrum of the input, it was necessary to specify a time window. With the wave propagating, however, it is possible for it to actually propagate out of the window. This occurs in Figure 2.4, but because of the periodicity inherent in the spectral analysis, the wave from the neighboring window propagates into view from the left. The lesson here is that the window size must be chosen not only to allow proper characterization of the spectral content, but also to allow room for propagating the signal. Consequently, there is a connection between the size of the time window and the allowable distance the wave can move. A point of interest that may seem odd at first, is that the window size is determined by the slowest component, not the fastest.

Propagating Dispersive Signals

Let the initial shape and transfer function be the same as in the previous example but choose the spectrum relation to be

$$G_n = e^{-ik_n x}, \qquad k_n = \sqrt{\omega_n \omega_o}/c_o$$

where ω_o is a constant. (Strictly speaking the transfer function should also include the damped vibration terms, but the present form is adequate for far predictions.) The amplitude spectrum is obviously as given before but the reconstructions shown in Figure 2.6 are very different. What starts out as a well-defined triangular shape disperses as it propagates. Not only does it change its shape, but it is no longer possible to identify the so-called "pulse" that is propagating. This is characteristic of dispersive signals and is due to the relation between k and ω being nonlinear. If the speed is slow for low frequencies, then it takes a very long time for the low-frequency part of the signal to arrive. Thus it will have a long tail.

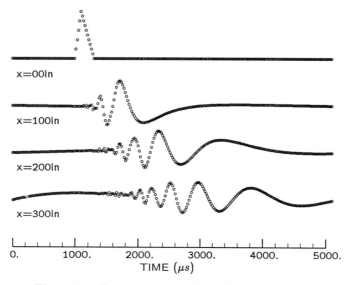

Figure 2.6: Reconstructions for a dispersive wave.

Figure 2.6 also shows the results of propagating a dispersive signal a large distance. The curious feature is that it is the lower frequencies that propagate from the left into view first. To minimize the effect of this it is a good idea to pad half the window with zeros. Thus the creeping components must first traverse the zeros before they come into view. Since the higher frequencies usually arrive sooner than unwanted lower frequencies, then (depending on the circumstance) the unwanted components can often simply be filtered out. Reference [63] shows two ways of doing this. The first method is simply by band-pass filtering, i.e., everything below a certain frequency is blocked. The second scheme uses an n-point moving average but retains the difference and not the average. This latter scheme can do an excellent job in reconstructing the propagating wave. The most effective scheme, however, is simply to increase the window size by adding more terms. This is feasible because of the efficiency of the FFT algorithm and its essentially linear dependence on N.

Integration of Signals

Inherent in the spectral analysis approach is the assumption that all functions of interest can be both integrated and differentiated. However, during the integration stage care must be taken to include the proper constants of integration. This is of special importance when, for example, obtaining displacement from strain since some unexpected side effects arise.

To illustrate some of the problems that arise due to the enforced periodicity of the results, Figure 2.7 shows the displacement at the impact site of a beam. The periodicity requires the beginning and final points of the reconstruction to be the same. But unlike the strain-from-force case there is also the effect due to an unaccounted for coefficient of integration (since displacement is obtained by double integration from strain). This figure shows the profound effect of the window size. For this particular beam problem a non-zero value of displacement should persist for all time after the passage of the pulse. However, periodicity of the result on the finite window forces the end displacement to zero. Actually, the trace is completely continuous from the end of the window to the beginning of the next.

This problem is minimized when the output is at the same differentiation level as the input. For example, when obtaining strain at one location from strain at another, then padding with zeros can be used. But in the case of strain from force it is seen that padding does not help. What is required is a window size large enough so that the slope of the trailing edge of the history is not significant. For a given sample rate, the window can only be increased by increasing the number of points in the transform. Conversely, for a fixed number of points, the sample rate ΔT must be increased, thus distorting the higher frequencies and running the risk of aliasing.

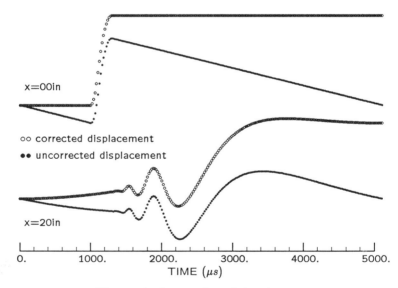

Figure 2.7: Integration of signals.

To understand the effect in time, it is of interest to write the differential relations in terms of the frequency and wave number for both beams and rods. (More details on the origin of the rod and beam equations can be found in Chapters 3 & 4.) The simplest propagating solutions for rods and beams are, respectively

$$\text{ROD:} \quad u(x,t) \;=\; \sum A_n e^{-i(kx - \omega t)}, \qquad k = a\omega$$

$$\text{BEAM:} \quad v(x,t) \;=\; \sum B_n e^{-i(kx - \omega t)}, \qquad k = a\sqrt{\omega}$$

A table can now be constructed that shows the interrelationship of all the derivatives

ROD   BEAM

$$\frac{dv}{dx} \to kv$$

$$\frac{du}{dx} \to ku \to \omega u \to \frac{du}{dt} \qquad\qquad \frac{d^2 v}{dx^2} \to k^2 v \to \omega v \to \frac{dv}{dt}$$

$$\frac{d^3 v}{dx^3} \to k^3 v$$

$$\frac{d^2 u}{dx^2} \to k^2 u \to \omega^2 u \to \frac{d^2 u}{dt^2} \qquad \frac{d^4 v}{dx^4} \to k^4 v \to \omega^2 v \to \frac{d^2 v}{dt^2}$$

Thus strain and velocity are at comparable levels of ω, and likewise for the loading and accelerations. This chart can be interpreted by noting, for example, that for beams integration of strain twice with respect to x to find displacement is equivalent to integration of velocity once with respect to time. The displacement is known, therefore, to within a linear function of time

$$\hat{v} = \int\!\!\int \hat{e}\, dxdx + C_1 t + C_2$$

There are many ways of establishing these coefficients of integration. A simple method is to curve fit the first portion of the signal with a straight line and subtract it from the total signal. In fact, it is good practice always to include an initial zero portion of the signal so as to be able to determine if there is an integration problem. Perhaps the simplest procedure is always to deal with signals that are nearly stationary. This is achieved by using time rates of the quantities of interest. Then after reconstruction the signal can be time integrated in the time domain. This is what was done to obtain the corrected displacements of Figure 2.7. The corresponding displacement rates are shown in Figure 2.8.

It is worth pointing out that this integration problem also arises when accelerometers are used to predict the strain or displacement.

2.4   Wave Behavior of the Motion

The dynamic solution is comprised of the superposition of many harmonics and it was seen from before that some are waves, some are damped waves, while some

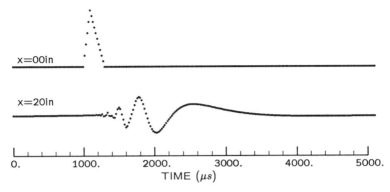

Figure 2.8: Displacement rates corresponding to Figure 2.7.

others are vibrations. All these combine to give the observed motion. It is of interest to know how to study the propagating part of the disturbance since that is what arrives at a remote location. It must be reiterated, however, that for all practical problems, all types of solutions are present, and it may be unwarranted to attempt to isolate the so-called "wave."

It was shown that in any local region of interest the solution can always be written in the exponential form

$$u(x,t) = \sum \hat{F}_n G_n(kx) e^{i\omega_n t} = \sum \hat{F}_n e^{-ik_n x} e^{i\omega_n t}$$

for a typical mode. This is valid even though the solution itself may not be of the exponential form (as in the Bessel solution, for instance). Thus it will be sufficient to study superpositions of this form.

Phase Speed

At a typical frequency, the wavenumber can be written in terms of its real and imaginary parts as

$$k = k_R + ik_I$$

giving the wave in the form

$$u(x,t) = \sum \hat{F}_n \, e^{-k_I x} e^{-i(k_R x - \omega t)}$$

This is recognized as having three parts: an amplitude spectrum $\hat{F}_n$, a spatially decaying term $e^{-k_I x}$, and the propagating sinusoids $e^{-i(k_R x - \omega t)}$. The phase speed of these sinusoids is given as

$$c = \frac{\omega}{k_R} \qquad\qquad (2.19)$$

This is the speed at which the individual harmonics move as depicted in Figure 2.1.

Group Speed

Since it is the behavior of the superposition of all the sinusoids that forms the observed signal, then it is of interest to investigate how this group response differs from the individual sinusoids. Consider the interaction of two neighboring propagating components, i.e.,

$$u(x,t) = \hat{F}_n\, e^{-ik_n x}\, e^{i\omega_n t} + \hat{F}_{n+1}\, e^{ik_{n+1} x}\, e^{i\omega_{n+1} t}$$

If these can be rewritten in terms of a central frequency $\omega^* = (\omega_n + \omega_{n+1})/2$, wavenumber $k^* = (k_n + k_{n+1})/2$, and amplitude $F^* = (F_n + F_{n+1})/2$, then the result is

$$u(x,t) = \hat{F}^* e^{-i(k^* x - \omega^* t)} 2\cos\left[\frac{\Delta k}{2}\left(x - \frac{d\omega^*}{dk^*}t\right)\right] \tag{2.20}$$

This resulting wave is comprised of two parts besides the amplitude spectrum. There is a sinusoid (called the carrier wave) of average frequency ω^* and wavenumber k^*, and travels with average speed $c^* = \omega^*/k^*$. But this is modulated by another wave (called the group wave) of wave number $\frac{1}{2}\Delta k$ and frequency $\frac{1}{2}\Delta k\, d\omega^*/dk^*$ and travels at a wave speed of $d\omega^*/dk^*$. The phase velocity of the modulation is called the group speed. That is,

$$c_g \equiv \frac{d\omega}{dk} = c/(1 - k\frac{dc}{d\omega}) = c + k\frac{dc}{dk} \tag{2.21}$$

In general, the group speed is numerically different from the phase speed. To amplify on this, consider someone moving at the group speed, then

- $c > c_g$ means the carrier waves appear to originate at the rear of the group, move to the front and then disappear.

- $c = c_g$ means there is no relative motion of group and carrier.

- $c < c_g$ mean the waves seem to originate at the front of the group, move to rear and then disappear.

While the above analysis is for only two harmonics, it can be imagined that many harmonics interact in a similar manner giving rise to a carrier wave modulated by a group wave. Then, however, it is not possible to "see" the individual sinusoids — only the group superposition is observed. It cannot be emphasized enough the importance of the group speed behavior because this is what is actually there. The ability to synthesize or assemble the group behavior is a unique feature of spectral analysis coupled with the FFT algorithm. Put another way, the FFT allows both the assembling and disassembling of the group response.

Some Examples

Recall that the dispersion relation for the simple wave equation is

$$k_R = k = a\omega$$

Therefore the phase and group speeds are

$$c = \frac{\omega}{k} = 1/a, \qquad c_g = \frac{d\omega}{dk} = 1/a = c$$

For non-dispersive systems the phase and group speeds are the same and this is the key to their simple descriptions. On the other hand, the dispersion relation for the beam is

$$k^2 = a\omega$$

giving the two speeds as

$$c = k/a = \sqrt{\omega/a}, \qquad c_g = 2k/a = 2\sqrt{\omega/a} = 2c$$

The group speed is twice the phase speed. For dispersive systems the group and carrier phase speeds are not the same. This is why a pulse changes shape as it propagates.

It is possible to have very complicated group behavior. For example, it is possible to have a solution for which the group speed is positive (the disturbance moves forward) and yet the phase speed is negative. According to Reference [79], as a caterpillar moves forward, the ripples in its body move from head to tail.

Summary of Wave Relations

The wave for a particular mode can be written in any of the following forms

$$u(x,t) = \mathbf{A}e^{-ik(x-ct)} = \mathbf{A}e^{-i\frac{2\pi}{\lambda}(x-ct)} = \mathbf{A}e^{-i(kx-\omega t)}$$

The amplitude $\mathbf{A}$ can be complex and the above form is for a forward moving wave. Following is a collection of terms usually used in wave analysis

$$\phi \;=\; \text{phase of wave [radians]} \;=\; (kx - \omega t) = \frac{\omega}{c}(x - ct) = \frac{2\pi}{\lambda}(x - ct)$$

$$c \;=\; \text{phase velocity [length/time]} \;=\; \frac{\omega}{k} = 2\pi\omega\lambda$$

$$c_g \;=\; \text{group speed [length/time]} \;=\; \frac{d\omega}{dk}$$

$$\omega \;=\; \text{angular frequency [radians/time]}$$

$$f \;=\; \text{cyclic frequency [hertz, 1/time]} = \frac{\omega}{2\pi}$$

$$\lambda \;=\; \text{wavelength [length]} \;=\; 2\pi\frac{c}{\omega} = \frac{2\pi}{k}$$

$$T \;=\; \text{period [time]} \;=\; \frac{1}{f} = \frac{2\pi}{\omega}$$

$$k \;=\; \text{wavenumber [1/length]} \;=\; \frac{2\pi}{\lambda} = \frac{\omega}{c}$$

It should be kept in mind, however, that a general disturbance will include non-propagating and attenuated components also.

2.5   Experimental Aspects of Wave Propagation

This section summarizes some of the experimental aspects of recording the wave signals in structures. By necessity it is only cursory, but it is hoped that it gives a feel for the type and quality of data recorded.

Characteristics of Wave Testing

Testing associated with waves is somewhat different from that of vibrations. Some of the characteristics of the testing to record transient signals associated with impact and stress wave propagation in structures are:

- High frequency content (1 kHz - 1 MHz).

- Initial portion of the signal is very important.

- Precise triggering is often difficult.

In contrast to vibration-type problems, this testing has the further characteristics:

- Only a small number of "runs" are performed.

- The number of channels used are usually limited (2 - 4).

- Only a short duration of the trace is analyzed (0 - 2000 μs).

- For proper analysis, the data must be in digital form.

The advantage of digitizing a signal is that thereafter it does not lose fidelity and all subsequent transport of it does not deteriorate its quality. However, it must be kept in mind that distortion can be introduced during the initial stage of converting the analog signal into digital form. This is something that must be monitored carefully lest spurious frequency components be generated.

Measurement Techniques

There is quite a range of measurement techniques available for studying wave propagation problems. Of course, some are more suitable for certain situations than
others, but since the emphasis in this book is on structures, then only those techniques most commonly used for structural analysis will be surveyed.

Most types of electrical resistance strain gages are usually adequate and since
the events are of short duration, temperature compensation is usually not an issue.
The largest size gage usable is dependent on the highest significant frequency (or
shortest wavelength) of the signal, but generally gages equal to or less than 3 mm
(0.125 in) are good for most situations. A rule of thumb for estimating the maximum
length is

$$L = \tfrac{1}{10} c_g / f \qquad (2.22)$$

where L is the gage length, f is the highest significant frequency, and c_g is the group
speed. An example of a typical gage is Micro Measurements®: ED-DY-125AD-350
and according to the above formula is good to about 160 kHz. Other aspects of
the use of strain gages in dynamic situations are covered in References [174,22].
Either constant-current, potentiometer or wheatstone bridge gage circuits can be
used. The simplest to use is the constant- current source, but a cheaper and more
reliable approach is to use 2 or 4 rechargeable batteries in a potentiometer circuit.
Nulling of the wheatstone bridge is not necessary for dynamic problems since only
the a.c. portion of the signal is recorded.

A wide variety of small accelerometers are available. An example of the one used
in Reference [230] is the Endevco® type M75 accelerometer. There is a problem
with accelerometers, however, in that if the frequency is high enough, ringing will
be observed in the signal. This can often occur for frequencies as low as 20 kHz.
Generally speaking, accelerometers do not have the frequency response of strain
gages, but they are movable, thus making them more suitable for larger, complex
structures. Also, note that if the stress is of interest, then integration of the signal
is required.

An entirely different approach is to use fiber optics as a displacement transducer.
The basic idea behind using fiber optic bundles is that some of the fibers project
the light onto the structure, while the rest of them collect the reflected light and
bring it to a photometer. Obviously the closer the object the more light the fibers
collect and therefore the photometer reading can be calibrated to read proximity.
Because photometers can have very short rise times, frequency response is excellent.
A commercial version found to be useful is the MTI® Fotonic sensor. From the
spectral estimation point of view, displacement is not the best signal to be dealing
with — it is most desirable to be dealing with nearly stationary signals. Thus to
use fiber optics effectively, it is necessary to convert to velocity by differentiation as
was done in Reference [62]. A final point about the use of fiber optics is that they
are particularly useful for determining phase relationships because then calibration

(which can be a problem) is not required.

Some very early wave monitors were of the capacitance gage type and Reference [49] gives a clear description of their construction. While this gage can have very high resolution it seems not to be very popular at present. Electro-magnetic-induction devices for measuring velocity are demonstrated in References [78,194].

Recording Techniques

There are many recording devices available commercially, and it would be impossible to survey them all. So, instead, some of the factors that go into choosing a recording setup will be given.

Reference [61] discusses what is the minimum set-up necessary for waveform recording and the following are the pieces of equipment that seem essential as a minimum. Good preamplifiers are essential for dynamic work. They not only provide the gain for the low voltages from the bridge but also perform signal conditioning. The former gives the flexibility in choice of gage circuit as well as reducing the burden on the recorder for providing amplification. A separate preamplifier is needed for each strain gage and an example of a typical one is the Tektronics® AM502 differential amplifier. This has 1 to 100k gain, d.c. to 1 MHz frequency response, and selectable band pass filtering. Generally, the analog filtering (which is always necessary) is performed at or below the Nyquist frequency associated with the digitizing. Historically, oscilloscopes were the primary means of recording (and displaying) the waveform. An oscilloscope with storage capability (the traces are stored on the phosphorus screen) is necessary for rapid set up to record transient events. A hard copy is made of the trace by taking a polaroid photograph of the image on the phosphorus screen.

This minimum set-up for recording waveforms has a few serious inadequacies. The primary ones are:

- **Triggering:** Its inflexible triggering often results in losing the initial portion of the trace. Having to use external or pretriggering with delay can be very cumbersome.

- **Quality:** While the electronics of the oscilloscope can be very good, often the quality of the results reduces to the quality of the blurred phosphorus image.

- **Data Transfer:** Numerical values are obtained by taking measurements off the polaroid photograph and this can be a tedious and inaccurate task.

The following are some additional pieces of equipment that, while not essential, certainly make the job easier, and in some cases can improve the quality of the data.

The recorded data need (eventually) to be uploaded to a computer. A convenient way of doing this is through a digitizing pad. This allows rapid digitizing of the polaroid photograph and is a significant improvement to the minimum set-up. An example is the SUMMAGRAPHICS®, Bitpad One, which has a 16 x 16 in. workspace, and 200 pt/in. resolution. This has an RS-232 port for easy communication with a computer. Having to manually run the cursor over the trace is not as big a hardship as it may seem, primarily because for dynamic work only a limited number of traces are analyzed anyway. But doing it this way, actually, has two side advantages. First, adjustments to the signal (such as removing reflections or significant noise) can be accomplished before digitizing. Secondly, once the data are in the computer, post-processing techniques such as interpolation, smoothing, etc., can allow a fairly good reconstruction of the signal from the limited number of input points. The digitizing pad can also be used to append signals to the recorded trace.

The most significant advance in recent years is the availability of digital recorders. This recorder replaces the phosphorus screen in the minimum set-up. An example is the Gould® Biomation 2805, Digital Waveform Recorder. This has 2 channels, 2048 8 bit words of memory, and a sample rate of 0.2 μs to 0.1 s. It is common for high-speed recorders to digitize to 8 bit words (although newer ones use 12 bit words.) This means it has a resolution of $1/256$ of the full scale. If the signal can be made to fill at least half full scale, then this resolution of about 1% of the maximum signal is adequate. This situation is best achieved by having good preamplifiers and not by requiring a wider voltage range on the digital recorder. Many of the newer digital recorders combine the features of an oscilloscope with those of a small stand-alone microcomputer. But, generally speaking, if the data are to be eventually transferred to a computer for analysis, then that is where all the processing should be done. However, the pre-processing capability of the recording machine is very useful during the setup stage. An example of this type of recorder is the Norland® 3001 digital oscilloscope with 3217A plug-ins. This has 4 channels, 4096 x 8bit words of memory and a sample rate of 0.1 μs to 7 ms. The basic characteristics of the Norland are about the same as for the Biomation; the special features are just more convenient to use. The memory of 4096 samples per channel is more than enough required for most structural dynamics problems. The transfer of the data to a computer via the RS-232 connection is very simple and is initiated from the keyboard.

Finally a word about the computer. There is a lot of flexibility here and any computer along the lines of an IBM®-AT or compatible is more than adequate. This particular machine supports a hard disk drive, the 80287 math coprocessor and many expansion slots. It also has the ability to create a sizable virtual disk drive so that during data transfer no data should be lost. If such a computer is available, then many new opportunities for recording also present themselves since there are many A/D converters on the market that are inexpensive and plug right into the expansion slots.

2.6   Signal Processing and Spectral Estimation

It is necessary to record an infinitely long trace in order to exactly characterize a dispersive signal. This is obviously impractical, so the problem reduces to one of obtaining best estimates for the spectra from a finite trace. This section summarizes some of the necessary procedures used for spectral estimation. While some of these procedures are similar to those used for vibration signals, the differences are significant.

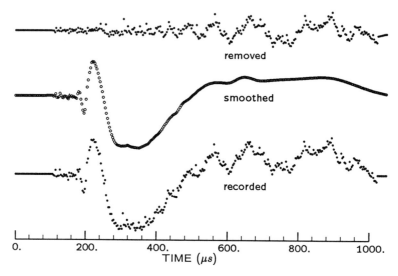

Figure 2.9: Effect of moving-average smoothing.

Smoothing

The measured signal can contain unwanted contributions from at least two sources. The first is the ever present electrical noise. Analog filters can be used to remove much of this, although care must be taken not to remove some of the signal itself. The second may arise from unwanted reflections. This is especially prevalent for dispersive signals as the recording is usually extended to capture as much as possible of the initial passage. There are many ways of smoothing digital data but it seems the simplest is just to use moving averages of various amounts and points of application. Figure 2.9 shows an example of smoothing a signal that has both noise and reflections present. The following sequence was used on data values sampled at every 1 μs :

5-point	starting at	0 μs
33-point	starting at	220 μs
77-point	starting at	450 μs
201-point	starting at	600 μs

The last averaging brings the signal to zero in the vicinity of 1000 μs. Also shown in the figure is that part of the signal removed; this can be monitored to detect if too much of the signal is being filtered.

Window

Another problem that arises when using a finite sample of an infinite signal is that of leakage. As shown in the previous chapter, the effect can be viewed as the product of two time-domain functions: the signal of interest and the window used. None of the windows usually used in vibration studies seem particularly appropriate for transient signals of the type obtained from propagating waves. For this reason, only the rectangular window is used (in conjunction with padding and appending) in the examples to follow. A problem with rapid truncation is that the reconstructed signals exhibit significant disturbances at the point of truncation. If the sample size is large compared to the length of interest, then this is not a serious problem because the disturbances can be easily identified for what they are and not be confused with the real signal.

It is also necessary to give the signal some room in which to propagate without components from neighboring windows coming into view. In fact, the maximum distance the wave can be propagated without components from neighboring windows drifting into view corresponds to the distance traveled by the lowest-frequency component for a time equal to the zero padding. This is estimated as

$$x < T_{\text{pad}} \; c_g \tag{2.23}$$

where T is the sampling window and c_g is the group speed of the lowest frequency component.

Appending

It usually happens that a sufficient length of trace is not available to simulate an infinite record or that reflections have become a problem. If it is not adequate to just put an averaged line through the reflections, then the signal can be appended by making use of the usually available long-term theoretical solution. For beams, for example, at a large time compared to the duration of the input pulse, the strain is given by

$$e(x,t) \approx \frac{A}{\sqrt{t}} \sin\left[\frac{x^2}{Bt} - \frac{\pi}{4}\right] \int_o^T F(\tau)d\tau \approx \frac{\text{constant}}{\sqrt{t}} \sin\left[\frac{x^2}{Bt} - \frac{\pi}{4}\right]$$

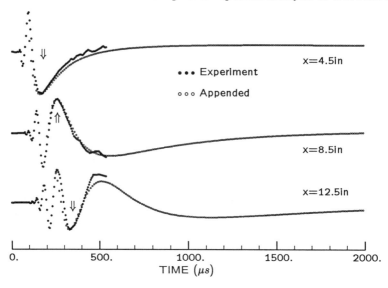

Figure 2.10: Appending begun at N = 1,2,3 indicated by the arrows.

This shows that the shape of the strain history becomes less and less dependent on the particular input history and only dependent on the impulse transferred. By tracking where the maxima occur, then the following appending formula can be obtained

$$e(t) = e(t_o)\sqrt{\frac{t_o}{t}} \sin\left[(4N - 1)\frac{t_o}{t} - \frac{\pi}{4}\right] \qquad (2.24)$$

where $e(t_o)$ is the strain at the Nth maximum. Similar expressions can be established for plates and rods.

Figure 2.10 shows examples of appended signals compared to the experimentally recorded. The arrows indicate where the matching is imposed. It is apparent that the appending is best when done on the last phase portion of the history. When the two strain histories are matched, the appending will have most effect on the low-frequency components. If use is made of appending, then it appears that small signal lengths (which are rich in the high-frequency content) can give adequate estimates of the complete spectra. It should also be borne in mind that for forward propagation studies the low frequencies play a minor role anyway. It is only when propagating backward (when reconstructing the input force history, for example) that this is important.

The length of signal necessary depends on where it is sampled. A larger signal length is required at the non-zero x position.

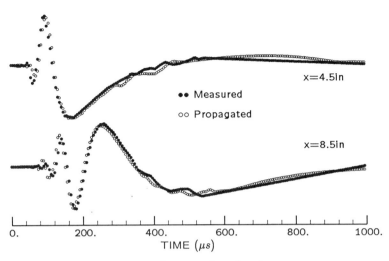

Figure 2.11: Propagating signals.

Propagation and Reconstruction of Signals

Once the signal spectrum has been accurately estimated, then it can be manipulated in a straightforward manner to obtain any of the quantities of interest. In particular, it can be propagated forward for predictive purposes or backward for identification purposes.

As a check on the consistency of the strain predictions, it is useful to reconstruct the strains at each location using the gage data at the other location. The predictions for the beam data of Figure 2.10, for example, shown in Figure 2.11, are quite good.

As regards spectral estimation from experimental data, if two measurements are made, then this approach can be used to perform averaging of them and thus obtain a better estimate.

Chapter 3

Longitudinal Waves in Rods

Rods and struts are very important structural elements and form the basis of many truss and grid frameworks. Since their load bearing capability is axial, then as waveguides they conduct only longitudinal wave motion.

This chapter begins by using elementary mechanics to obtain the equations of motion of the rod. As it happens, the waves are governed by the simple wave equation and so many of the results are easily interpreted. The richness and versatility of the spectral approach is illustrated by considering such usually complicating factors as viscoelasticity and non-uniform dimensions. Particular emphasis is placed on the waves interacting with discontinuities such as boundaries and changes in cross-section. The treatment of the very important topic of multiple modes is introduced and illustrated by way of a coupled thermoelastic problem. The chapter concludes by introducing some of the higher-order rod theories. In all the problems considered, the essential quantity to determine is the transfer function and the form of this is elaborated on.

All the examples of this chapter were run on a PC computer using the program RODCOMP listed in Appendix D.

3.1 Elementary Rod Theory

There are various schemes for deriving the equations of motion for structural elements. The approach taken here is to begin with the simplest available model, and then as the need arises to append modifications to it. In this way, both the mechanics and the wave phenomena can be focused on without being unduly hindered by some cumbersome mathematics.

The elementary theory considers the rod to be long and slender, and assumes it supports only one-dimensional axial stress. It further assumes that the lateral contraction (or the Poisson's ratio effect) can be neglected. Both of these restrictions will be removed later in the chapter when the Mindlin-Herrmann rod theory is developed.

Equation of Motion

Let $q(x,t)$ be the externally applied body force per unit volume and $u(x,t)$ be the displacement in the x direction. Then, with reference to Figure 3.1, the balance of forces gives

Figure 3.1: Element of rod with loads.

$$-F + [F + \Delta F] + qA\Delta x = \rho A \Delta x \ddot{u}$$

where ρA is the mass density per unit length of the rod. If the Δ quantities are very small, then the equation of motion becomes

$$\frac{\partial F}{\partial x} = \rho A \frac{\partial^2 u}{\partial t^2} - qA \tag{3.1}$$

The independent variables are x and t. It is now desirable to write the equation only in terms of the displacement. To do this, first consider the strain-displacement relation

$$e = \frac{\partial u}{\partial x} \tag{3.2}$$

and assume the material behavior to be linear elastic in the 1-D form

$$\frac{F}{A} = \sigma = Ee, \qquad E = \text{Young's Modulus} \tag{3.3}$$

Then combining with the above gives

$$\frac{\partial}{\partial x}\left\{ EA\frac{\partial u}{\partial x} \right\} = \rho A \frac{\partial^2 u}{\partial t^2} - qA \tag{3.4}$$

If the stress (or strain) is taken as the dependent variable, it also would have an equation of similar form. That is, all dependent variables have an equation of the form

$$c_o^2 \frac{\partial^2 u}{\partial x^2} - \frac{\partial^2 u}{\partial t^2} = 0, \qquad c_o = \sqrt{\frac{EA}{\rho A}} \tag{3.5}$$

for the homogeneous part (if the section and material properties are constant). The waves in the rod are governed by the simple wave equation and therefore the general solution is that of D'Alembert, given by

$$u(x,t) = f(x - c_o t) + F(x + c_o t) \tag{3.6}$$

This solution does not hold for the general rods of interest in this chapter and therefore will not be pursued any further in its present form.

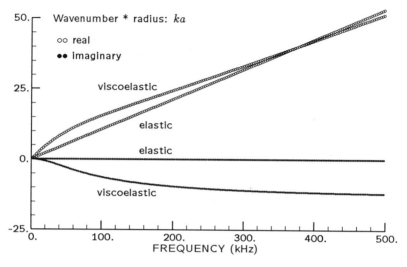

Figure 3.2: Spectrum relations for a rod.

Spectral Analysis

Let the solution have the spectral representation developed in the last chapter. The kernel solutions corresponding to this are obtained by considering the homogeneous equation

$$\frac{d}{dx}\left\{EA\frac{d\hat{u}}{dx}\right\} + \omega^2\rho A\hat{u} = -\hat{q}A \tag{3.7}$$

This is the basic equation used in the remainder of the chapter. Assume that both the modulus and area do not vary with position, then the homogeneous differential equation for the Fourier coefficients becomes

$$EA\frac{d^2\hat{u}}{dx^2} + \omega^2\rho A\hat{u} = 0$$

Since this equation has constant coefficients, then it has the obvious solution

$$\hat{u}(x) = \mathbf{A}e^{-ikx} + \mathbf{B}e^{+ikx}, \qquad k = \omega\sqrt{\frac{\rho A}{EA}} \tag{3.8}$$

where $\mathbf{A}$ and $\mathbf{B}$ are the undetermined amplitudes at each frequency. When combined with the time variation, this corresponds to two waves: a forward moving wave and a backward moving wave. That is,

$$u(x,t) = \sum \mathbf{A}e^{-i(kx-\omega t)} + \sum \mathbf{B}e^{+i(kx+\omega t)} \tag{3.9}$$

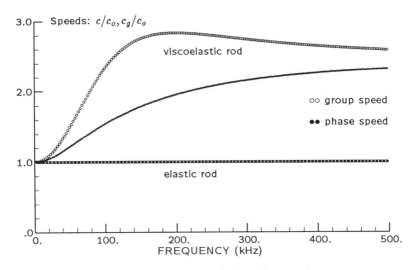

Figure 3.3: Dispersion relations for a rod.

As usual, it is understood that all quantities $(\mathbf{A}, \mathbf{B}, k,$ etc.$)$ depend on the frequency ω_n.

The spectrum relation for this case is simply

$$k = \pm \omega \sqrt{\frac{\rho A}{E A}} \qquad (3.10)$$

which is linear in frequency. This gives, for the dispersion relation, constant phase and group speeds of

$$c = \frac{\omega}{k} = \sqrt{\frac{E A}{\rho A}} = c_o , \qquad c_g = \frac{d\omega}{dk} = \sqrt{\frac{E A}{\rho A}} = c_o \qquad (3.11)$$

This simple behavior is shown in Figures 3.2 and 3.3, where it is seen that the phase and group speeds are just constant. Also shown are some other complicated behaviors as a reminder that even the simple rod is capable of a wide variety of behaviors and the goal of this chapter is to develop the tools necessary for their description.

As a useful reference, the properties and speeds of some typical structural materials are given in Table 3.1.

3.2 Basic Solution for Waves in Rods

Once the spectral approach is adopted for rods, then many familiar features will appear in a slightly different guise. The purpose of this section is to collect many

Material	Modulus		Density		Wave Speed	
	msi	GPa	lb/in.^{3}	kg/m^2	in./μs	km/s
Aluminum	10.6	72.7	0.00026	2.7	0.206	5.23
Brass	12.0	82.3	0.00079	8.1	0.135	3.43
Concrete	4.0	27.4	0.00023	2.4	0.132	3.35
Copper	13.0	89.1	0.00083	8.5	0.141	3.58
Epoxy	0.5	3.4	0.00010	1.0	0.70	1.78
Glass	10.0	68.6	0.00023	2.3	0.209	5.31
Rock	4.4	30.1	0.00024	2.5	0.140	3.56
Steel	30.0	206	0.00073	7.5	0.203	5.15

Table 3.1: Typical properties for some engineering materials.

of these together and show their interrelation. This, it is hoped, will make the transition to the full spectral analysis of the later chapters more accessible.

Relation to the D'Alembert Solution

The phase term in the solution (3.9) can be written as

$$\pm kx + \omega t = \frac{\omega}{c_o}(\pm x + c_o t) = \omega \xi \quad \text{or} \quad \omega \eta \tag{3.12}$$

where $\xi \equiv x + c_o t$ and $\eta \equiv x - c_o t$, giving

$$u(x,t) = \sum A e^{i\omega \eta} + \sum B e^{i\omega \xi}$$

These are recognized as discrete Fourier transforms with respect to η and ξ, respectively. Consequently, the solution can be written directly as

$$u(x,t) = f(\eta) + F(\xi) = f(x - c_o t) + F(x + c_o t)$$

where f, F are the inverse transforms of $\mathbf{A}, \mathbf{B}$, respectively. This, of course, is the D'Alembert solution and says that the pulse keeps the same shape as it propagates. Of course, to be able to write the solution in this form required that the relation between k and ω be linear. Such a case is not true in general and this result has limited value.

Stress and Force Behavior

Using the subscripts i and r to represent the forward and backward moving waves, respectively, then the other mechanical quantities (at a particular frequency) can be obtained as

Displacement: $\quad u_i = \mathbf{A}e^{-i(kx-\omega t)} \qquad u_r = \mathbf{B}e^{i(kx+\omega t)}$

Strain: $\qquad\quad\ e_i = -iku_i \qquad\qquad e_r = iku_r$

Stress: $\qquad\quad \sigma_i = -ikEu_i \qquad\quad F_r = ikEAu_r$

Force: $\qquad\quad F_i = -ikEAu_i \qquad F_r = ikEAu_r$

Velocity: $\qquad \dot{u}_i = i\omega u_i = ikc_o u_i \quad \dot{u}_i = i\omega u_r = ikc_o u_r$

giving the following interesting interrelationships

$$\dot{u}_i = -\frac{c_o}{E}\sigma_i = -\frac{c_o}{EA}F_i, \qquad \dot{u}_r = +\frac{c_o}{E}\sigma_r = +\frac{c_o}{EA}F_r \qquad (3.13)$$

Thus the profiles of the particle velocity and stress are the same. Notice that a tensile stress moving forward causes a negative velocity (i.e., the particles move backward.) Also, the convention for the sign of the forces is that of the stress.

Force History Applied at One End

The following problem of the imposition of a force history at one end of the rod is treated in detail as a preliminary to analyzing the case of waves interacting with discontinuities. The steps are laid out in more detail here than will be done in the later examples.

Let the end of the bar at $x = 0$ be subjected to a force history $F(t)$, that is,

$$\text{at } x = 0: \qquad EA\frac{\partial u(x,t)}{\partial x} = F(t)$$

This is put into the form for the present analysis by assuming that all functions of time can be written in the spectral form. Thus the boundary condition becomes in expanded form

$$EA\frac{d}{dx}\sum_n \hat{u}_n(x)e^{i\omega_n t} = \sum_n \hat{F}_n e^{i\omega_n t}$$

This has to be true for all time, hence the equality must be true on a term by term basis giving

$$EA\frac{d\hat{u}_n}{dx} = \hat{F}_n$$

The problem stated will always reduce to establishing the relationships at a particular n (or frequency ω_n), hence the subscript will usually be understood. The condition at $x = 0$ now becomes (using the solution (3.9) and since there is only a forward moving wave)

$$EA\{-ik\mathbf{A}\} = \hat{F} \qquad \text{or} \qquad \mathbf{A} = -\frac{\hat{F}}{ikEA}$$

The complete solution for the forward moving wave is therefore

$$u(x,t) = -\sum \frac{\hat{F}}{ikEA}e^{-i(kx-\omega t)}$$

or with every term written out explicitly

$$u(x,t) = -\frac{1}{EA}\sum_n \frac{\hat{F}_n}{ik_n}e^{-i(k_n x-\omega_n t)} \tag{3.14}$$

Knowing the force history, the displacement (and, consequently, the stress, etc.) can be determined at any location x. Examples of the resulting strain histories are shown in Figure 2.4. As is apparent, the disturbance travels at the now familiar constant speed of $c_o = \sqrt{EA/\rho A}$.

The transfer function for this case is

$$G(x,\omega) = -\frac{1}{EA}\frac{1}{ik_n}e^{-ik_n x}$$

The position x contributes the phase change. Note that the amplitude varies as $1/k$ and therefore also as $1/\omega$. The displacement is related to force by a single integration (that is why k occurs in the denominator). Its effect is to essentially amplify the low-frequency components and may cause windowing problems as discussed in Chapter 2. The transfer function for the velocity, on the other hand, is

$$G(x,\omega) = -\frac{c_o}{EA}e^{-ik_n x}$$

and the amplitude does not depend on frequency. Consequently, the windowing parameters used for transforming the force will also be appropriate for reconstructing the velocity.

3.3   Reflection from Boundaries

Once a wave is initiated, it will eventually meet an obstruction in the form of a boundary or discontinuity since all structures are finite in extent. Thus to extend the range of wave analysis, it is of great importance to know how the wave interacts with these obstructions. Some typical situations are considered here and in the next section.

Boundary Conditions

The incident wave generates a reflected wave in such a way that the two superpose at the boundary to satisfy the boundary conditions. The only waves that can be present are the two given by

$$u(x,t) = \sum \mathbf{A}e^{-i(kx-\omega t)} + \sum \mathbf{B}e^{+i(kx+\omega t)}$$

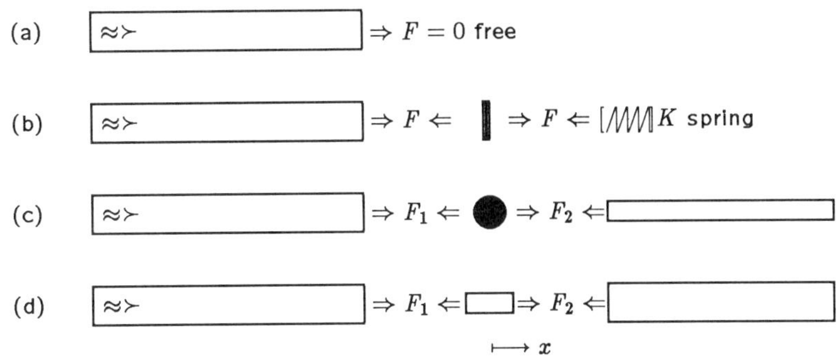

Figure 3.4: Free body diagrams for typical boundaries.

where $\mathbf{A}$ is associated with the known incident wave and $\mathbf{B}$ with the unknown reflected wave. This is an important point for later analyses: Since the differential equations establish all the possible solutions, then the number of choices is automatically prescribed — there is no possibility of "leaving something out." (This, of course, assumes that the choice of structural model is adequate for the problem.) The boundary conditions are set in terms of

$$\text{Displacement:} \quad u \ (\text{or } \dot{u} \text{ etc.})$$
$$\text{Force:} \quad EA\frac{\partial u}{\partial x}$$

A typical set of boundaries along with their equations is shown in Table 3.2. Note how easily the time domain conditions are converted into conditions on the spectral components $\hat{u}$.

Reflection from a Free End

The boundary condition at the free end $(x = 0)$ requires that the force be zero, that is, for each frequency component

$$EA\{-ik\mathbf{A} + ik\mathbf{B}\} = 0$$

This simply gives $\mathbf{B} = \mathbf{A}$, which says that the reflected displacement pulse is the same as the incident. By differentiating, it can be seen that the reflected stress pulse is inverted.

Reflection from an Elastic Boundary

Assume that at the end of the rod there is an elastic spring as shown in Figure 3.4b. The resistive force is proportional to displacement, hence at the boundary

BC Type	Imposed	Equation
Pinned	$u(0,t) = 0$	$\hat{u} = 0$
Free	$EA\dfrac{\partial u(0,t)}{\partial x} = 0$	$EA\dfrac{d\hat{u}(0)}{dx} = 0$
Spring	$EA\dfrac{\partial u(0,t)}{\partial x} = -K\,u(0,t)$	$EA\dfrac{d\hat{u}(0)}{dx} = -K\,\hat{u}(0)$
Dashpot	$EA\dfrac{\partial u}{\partial x} = -\eta\,\dfrac{\partial u(0,t)}{\partial t}$	$EA\dfrac{d\hat{u}(0)}{dx} = +\eta i\omega\hat{u}(0)$
Mass	$EA\dfrac{\partial u(0,t)}{\partial x} = -m\,\dfrac{\partial^2 u(0,t)}{\partial t^2}$	$EA\dfrac{d\hat{u}(0)}{dx} = +m\,\omega^2\hat{u}(0)$

Table 3.2: Some typical boundary conditions for rods.

$x = 0 : F(t) = -K\,u(0,t)$ or

$$EA\{-ikA + ikB\} = -K\{A + B\}$$

giving the reflected wave amplitude as

$$B = \frac{ikEA - K}{ikEA + K}A \qquad\qquad (3.15)$$

Since the wavenumber k is related to frequency, then it is apparent that each frequency component is affected differently and that, in general, the amplitude ratio is complex. Figure 3.5 shows some reconstructions of the pulse after it is reflected from springs of different stiffness. The incident pulse is a tensile triangle in each case. The behavior is varied, and to best explain it, consider some of the extreme limits.

First, consider the limits on the spring behavior. If there is no spring, i.e., $K = 0$, then

$$B = A\,, \qquad u_r = u_i\,, \qquad \sigma_r = -\sigma_i$$

which, of course, is the free end condition. On the other hand, if the spring if very stiff, i.e., $K = \infty$, then

$$B = -A\,, \qquad u_r = -u_i\,, \qquad \sigma_r = \sigma_i$$

This gives the fixed end condition. The stress is inverted in the first case and the displacement in the other. These limiting results are independent of frequency and hence the pulse will retain the same shape.

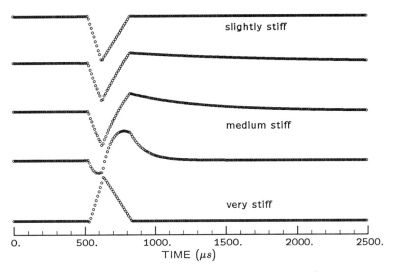

Figure 3.5: Reflected responses for various spring stiffnesses.

For a given spring stiffness, the limiting behavior on the frequency gives rise to the same limits as above. That is, the high frequencies experience a free end and the low frequencies experience a rigid end. Consequently, for a pulse with a spectrum of frequencies, the reflected signal is distorted as shown for the spring of medium stiffness. (Parenthetically, medium in this situation means relative to the amplitude spectrum of the wave.)

It is clear that for the medium stiffness spring there is a phase shift in the occurrence of the maximum strain. This is a typical example of how a nominally non-dispersive signal behaves dispersively when interacting with discontinuities. To amplify on this, consider the relation between the coefficients (the amplitude ratio) to be a transfer function. That is, $\mathbf{B} = G(\omega_n)\mathbf{A}$ and Figure 3.6 shows the plot. What is striking is that it has the form of a damped resonance, i.e., a peak in the imaginary part while the real part goes through zero. However, unlike resonance, the amplitude always remains unity and so there is no build up of amplitude, but there is the corresponding change in phase.

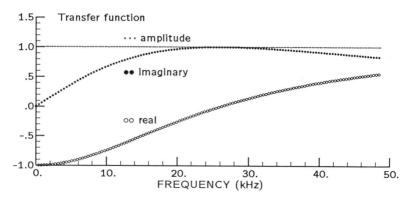

Figure 3.6: Transfer function for a spring end condition.

3.4   Reflections and Transmissions

Any change in cross section or material properties will cause the generation of new waves. While the actual situation is very complicated, in the present one-dimensional analysis only a longitudinal transmitted wave and a longitudinal reflected wave are generated. That is, the displacements in the two sections of the rod are taken as

$$\begin{aligned}
\hat{u}_1 &= \mathbf{A}_1 e^{-ik_1x} + \mathbf{B}_1 e^{ik_1x} \\
\hat{u}_2 &= \mathbf{A}_2 e^{-ik_2x}
\end{aligned} \qquad (3.16)$$

The analysis is performed by imposing continuity of force and displacement across the discontinuity.

More complicated problems, such as waves in curved rods (References [43,98,99] and [100,105,213]), and waves in non-collinear rods (References [12,54,135] and [147,235,237]), could also be treated, but these generally create new wave modes (such as flexural waves) and so will be left to the next two chapters.

Concentrated Mass

Consider two similar rod sections connected by a concentrated mass as shown in Figure 3.4c. The equation of motion of the mass and the continuity condition gives

$$\begin{aligned}
F_1 - F_2 &= m_J \ddot{u}_J \\
u_1 = u_2 &= u_J
\end{aligned}$$

where the subscript J refers to the joint. These become (since there is only a forward moving wave in rod 2 and both rod segments are similar)

$$EA\{-\mathbf{A}_1 + \mathbf{B}_1\}(ik) - EA\{-\mathbf{A}_2\}(ik) = m_J\{\mathbf{A}_2\}(i\omega)^2$$

$$A_1 + B_1 \; = \; A_2$$

Solving gives the amplitudes of the two generated waves as

$$B_1 = \frac{-m_J\omega^2}{2ikEA + m_J\omega^2}A_1, \qquad A_2 = \frac{2ikEA}{2ikEA + m_J\omega^2}A_1 \qquad (3.17)$$

These relationships are complex and frequency dependent; thus distortion of the wave can be expected. That is, the mass acts as a frequency filter. For the low frequencies, for example,

$$B_1 = 0, \qquad A_2 = A_1$$

showing the signal is unaffected. But at high frequencies

$$B_1 = -A_1, \qquad A_2 = 0$$

showing that the mass acts as a rigid end and does not transmit any of the wave. Notice that the same result is effected by increasing the size of the mass. The complete range of behavior of the transfer function for the transmitted and reflected waves are shown in Figure 3.7. Note that the sum of the two always adds up to unity.

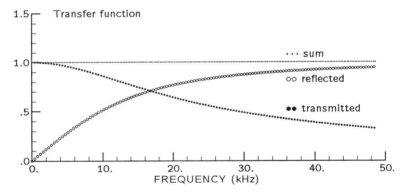

Figure 3.7: Transfer function amplitudes for a concentrated mass connection.

Stepped Rod

The case of transmission of waves from one rod to another is handled the same way as above. Balance of force at the (assumed massless) joint of Figure 3.4c and continuity of displacement gives

$$E_1 A_1 (-A_1 + B_1)(ik_1) \; = \; E_2 A_2 (-A_2)(ik_2)$$
$$A_1 + B_1 \; = \; A_2$$

Solving for the reflected and the transmitted coefficients then gives

$$\mathbf{B_1} = \left(\frac{1 - \sqrt{r_S r_D}}{1 + \sqrt{r_S r_D}}\right) \mathbf{A_1}, \qquad \mathbf{A_2} = \left(\frac{2}{1 + \sqrt{r_S r_D}}\right) \mathbf{A_1} \qquad (3.18)$$

where the following notations are used

$$r_S \equiv \frac{E_2 A_2}{E_1 A_1}, \qquad r_D \equiv \frac{\rho_2 A_2}{\rho_1 A_1}$$

Thus the response is independent of frequency. This is interesting in comparison to the case of the elastic spring. While the second rod and a spring behave the same statically, the mass distribution of the rod makes a profound difference dynamically. The expressions for the stresses are simply

$$\sigma_r = \frac{\sqrt{r_S r_D} - 1}{\sqrt{r_S r_D} + 1} \sigma_i, \qquad \sigma_t = 2\frac{\sqrt{r_S r_D}}{\sqrt{r_S r_D} + 1} \frac{A_1}{A_2} \sigma_i \qquad (3.19)$$

Reference [195] gives some experimental validation of these stepped rod results.

An interesting question is that since letting the modulus, area, or density go to zero, all intuitively indicating the situation of "no second rod," then how should the special limiting case of a free end condition be obtained? To answer this, recall that the balance of forces in the derivation of the differential equation gave

$$EA\frac{\partial^2 u}{\partial x^2} - \rho A\frac{\partial^2 u}{\partial t^2} = 0$$

That is, the wave behavior depends on the rod properties as specified in terms of the density ρA and the stiffness EA. Hence the solution can be written using only ratios of these, i.e., r_S and r_D. Looking at the limiting conditions, it is necessary to impose, instead, the limiting conditions on stiffness and density. That is, for a free end $r_S = 0($ or $r_D = 0)$ giving

$$\sigma_r = -\sigma_i, \qquad \sigma_t = 0$$

Similarly, for a fixed end impose $r_S = \infty$ (or $r_D = \infty$) giving

$$\sigma_r = \sigma_i, \qquad \sigma_t = 2\frac{A_1}{A_2}\sigma_i$$

Of course, these are the proper results. It is clear that the areas should not go to zero because of the stresses. Note that for the fixed end condition, there is a significant transmitted stress.

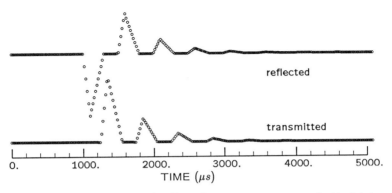

Figure 3.8: Transmitted and reflected responses near an elastic joint.

Reflections and Transmissions at an Elastic Joint

Consider two long bars with an insert of a different material or cross-section sandwiched between them as shown in Figure 3.4d. When the two long portions are of similar material, then this is essentially a split Hopkinson pressure bar and References [80,109] give some analysis of its use. Of interest here, however, is the role played by a finite rod in affecting the wave response and how that is reflected in the transfer function.

The waves in each section are represented as

$$
\begin{aligned}
\hat{u}_1 &= \mathbf{A}_1 e^{-ik_1 x} + \mathbf{B}_1 e^{ik_1 x} \\
\hat{u}_2 &= \mathbf{A}_2 e^{-ik_2 x} + \mathbf{B}_2 e^{ik_2 x} \\
\hat{u}_3 &= \mathbf{A}_3 e^{-ik_3 x}
\end{aligned}
\tag{3.20}
$$

The incident wave is $\mathbf{A}_1$ and there is only one wave in the third rod. Note that the rod forming the joint will have a standing wave since both forward and backward moving waves will be present simultaneously in the finite length. This is fundamentally different from the two waves in the semi-infinite first rod as will be seen. The boundary conditions at the joint interfaces, $x = 0$ and L, are

$$
\begin{aligned}
\text{at } x = 0: & \quad u_1 = u_2, & F_1 = F_2 \\
\text{at } x = L: & \quad u_2 = u_3, & F_2 = F_3
\end{aligned}
$$

Imposing these four conditions gives the following system of equations

$$
\begin{bmatrix}
-1 & 1 & 1 & 0 \\
(kEA)_1 & (kEA)_2 & -(kEA)_2 & 0 \\
0 & e^{-ik_2 L} & e^{ik_2 L} & -e^{-ik_3 L} \\
0 & -(kEA)_2 e^{-ik_2 L} & (kEA)_2 e^{ik_2 L} & (kEA)_3 e^{-ik_3 L}
\end{bmatrix}
\begin{Bmatrix}
\mathbf{B}_1 \\
\mathbf{A}_2 \\
\mathbf{B}_2 \\
\mathbf{A}_3
\end{Bmatrix}
=
\begin{Bmatrix}
1 \\
(kEA)_1 \\
0 \\
0
\end{Bmatrix}
\mathbf{A}_1
\tag{3.21}
$$

While this can be solved explicitly it is best left as part of the algorithm for solving for the transfer function of the wave. Such a scheme is shown in the program RODCOMP in the Appendix. If all the bars are elastic, then this relation simplifies considerably, but leaving it in its present form allows, for example, a solution for the viscoelastic joint. Figure 3.8 shows some of the responses on both sides of the insert for an incident tensile triangular pulse. It is apparent that a pulse is propagating backward and forward in the joint. Further, each reflected and transmitted pulse is similar in shape but diminishing in amplitude. It appears as if there are multiple copies of the pulse generated.

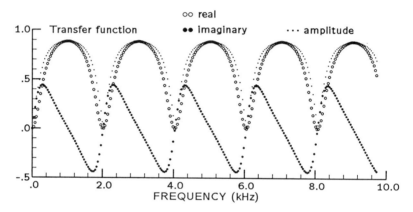

Figure 3.9: Transfer function for the reflection near an elastic joint.

In looking at the differences between the system (3.21) and the previous ones considered, the presence of the e^{ikL} terms are most significant. These are periodic functions, and, consequently, the solution will also express periodicity. Figure 3.9 shows an expansion of the transfer function at low frequencies. Its most striking feature is that it is indeed periodic. Also, as the real component goes through a maximum, the imaginary component goes through a zero. This is a resonance behavior, and unlike the elastic spring case, the amplitude also experiences a maximum. To summarize, the effect of the elastic joint on the response of each term is similar to that of a resonator. In the frequency domain this appears as peaks in the amplitudes of the transfer functions; in the time domain, it appears as repeated pulses.

In this particular example, each transmitted pulse is similar in shape, perhaps leading to the conclusion that the response is non-dispersive (i.e., "no change in shape"). But consider what happens when the joint becomes small — then the multiple transmissions overlap to give a shape that is definitely not triangular. The point is that when the full response is compared to the incident wave, then on this basis the elastic joint has a dispersive effect.

This raises another issue. The effect of the joint is to increase the total duration of the response: a single incident pulse of duration 100 μs, say, may become ten smaller pulses over a duration of 2000 μs. As a consequence, a time window chosen to describe the incident pulse may turn out to be too small to adequately contain the response. The resulting reflections or transmissions from neighboring windows will then propagate into view. These negative effects can be minimized by choosing a very large window with the consequent expense of a large N. An alternative approach is to add a little damping to the system so as to force the multiple reflections to zero sooner. This recommendation holds, in any event, because it also adds stability to the calculations, and (for what its worth) it makes the description more realistic since all real systems have some amount of damping. A scheme for adding this damping is discussed later in the chapter.

3.5   Coupled Thermoelastic Waves

Another generic type of loading that can give rise to stress waves is that of a thermal blast. While this is a very interesting problem in its own right (especially in the form of coupled thermoelastic waves), the treatment here is necessarily superficial. Its use is primarily as the first example to show the convenience of the spectral approach for coupled systems.

Governing Equations for Coupled Thermoelastic Waves

Following Reference [15] the coupled differential equations governing the axial displacement $u(x,t)$ and temperature change $T(x,t)$ are

$$\frac{\partial^2 u}{\partial x^2} - \frac{1}{c_o^2}\frac{\partial^2 u}{\partial t^2} = \alpha\frac{\partial T}{\partial x}, \qquad \frac{\partial^2 T}{\partial x^2} - \frac{1}{d_o}\frac{\partial}{\partial t} = \alpha\beta\frac{\partial^2 u}{\partial x \partial t} \qquad (3.22)$$

where d_o is the thermal diffusivity and α is the coefficient of thermal expansion. The parameter $\beta = E\alpha T_o/K$ (where K is the thermal conductivity) controls the amount of coupling between the two equations. Assume that both the displacement and temperature have spectral representations, then on substitution into the differential equations get

$$\frac{d^2\hat{u}}{dx^2} + \frac{\omega^2}{c_o^2}\hat{u} = \alpha\frac{d\hat{T}}{dx} \qquad (3.23)$$

$$\frac{d^2\hat{T}}{dx^2} - \frac{i\omega}{d_o}\hat{T} = i\omega\alpha\beta\frac{d\hat{u}}{dx}$$

These are ordinary differential equations with constant coefficients and have solutions of the form

$$\hat{u} = A e^{-ikx}, \qquad \hat{T} = \bar{A} e^{-ikx}$$

only if the following system of equations is satisfied

$$\begin{bmatrix} (-k^2 + (\omega/c_o)^2) & -i\alpha k \\ \omega\alpha\beta k & (-k^2 - i\omega/d_o) \end{bmatrix} \begin{Bmatrix} A \\ \bar{A} \end{Bmatrix} = 0 \qquad (3.24)$$

The only way to have a non-trivial solution for A and $\bar{A}$ is if the determinant of the matrix is zero. This is not so in general; however, if k is considered undetermined and adjusted in such a way as to force the determinant to zero, then a non-trivial solution is possible. (The procedure is analogous to that used in finding the eigenvalues of a matrix.) Multiplying the determinant out and imposing that it be zero allows k to be determined from the following characteristic equation

$$\left(k^2 - \frac{\omega^2}{c_o^2}\right)\left(k^2 + \frac{i\omega}{d_o}\right) + i\omega\alpha^2\beta k^2 = 0$$

This is quadratic in k^2 and therefore k itself can be easily determined as

$$\sqrt{2}k = \pm\sqrt{\left\{\left[\left(\frac{\omega}{c_o}\right)^2 - \frac{i\omega}{d_o}\right] - i\omega\beta\alpha^2\right\} \pm \left[\left(\frac{\omega}{c_o}\right)^2 + \frac{i\omega}{d_o}\right]\sqrt{1 - Q\beta\alpha^2}} \quad (3.25)$$

$$Q = \omega^2\left[2\frac{i\omega}{c_o^2} + \alpha^2\beta + \frac{2}{d_o}\right] \Bigg/ \left[\left(\frac{\omega}{c_o}\right)^2 + \frac{i\omega}{d_o}\right]^2$$

In all, there are four possibilities for the wavenumber (and consequently there are four modes of behavior) giving rise to a complete solution written as

$$\hat{u} = Ae^{-ik_1 z} + Be^{-ik_2 z} + Ce^{ik_1 z} + De^{ik_2 z} \qquad (3.26)$$
$$\hat{T} = \bar{A}e^{-ik_1 z} + \bar{B}e^{-ik_2 z} + \bar{C}e^{ik_1 z} + \bar{D}e^{ik_2 z}$$

But since $\hat{u}$ and $\hat{T}$ are coupled according to equation (3.24), then the amplitudes for each mode (the corresponding barred and unbarred coefficients) are related by

$$(\bar{}) = \frac{(-k^2 + \omega^2/c_o^2)}{i\alpha k}() \qquad (3.27)$$

where k for the appropriate mode must be used. Consequently, there are a total of four independent solution sets whose amplitude spectra are represented by the coefficients A, B, C, D. These are determined by the boundary and radiation conditions.

The Spectrum Relation

It turns out that, for aluminum say, the spectrum relations are very insensitive to the coupling parameter β (changes of many orders of magnitude are necessary before an effect is noticeable). Details of these and the following studies can be found in Reference [72]. The spectrum relations are dominated by the uncoupled case $(\beta = 0)$ giving approximately

$$k = \pm \frac{\omega}{c_o} \qquad \text{and} \qquad k = \pm \sqrt{\frac{i\omega}{d_o}}$$

The first set are obviously those of the longitudinal stress wave in the rod, the other is associated with the temperature change according to Fourier's Law. It is emphasized, though, that all modes are necessary to describe the displacement response and all modes are necessary to describe the temperature response. That is, aside from coupling of the differential equations there can also be coupling of the initial (or boundary) conditions, and all modes are necessary to satisfy these properly.

The mode 1 behavior has a negligible imaginary component and a nearly linear relation between frequency and the real part. Consequently, no attenuation (in space) is expected. Further, since the group speed is constant, then this portion of the response will propagate unchanged in shape. The mode 2 behavior is quite different in each regard. First, there is a sizable imaginary component indicating attenuation. Secondly, since the relation is nonlinear in frequency, that portion that is propagated will change its shape. Finally, in comparison to the first mode, the magnitudes of the respective wavenumbers are vastly different, indeed, nearly in a ratio of 100 to 1. Consequently, propagation speeds are exceedingly slow and attenuation will be that much more.

In summary, then, it can be said that irrespective of the particular initial boundary value problem, the mode 2 contribution to the displacement and temperature will be highly localized in space, whereas the other mode will send out a propagating non-dispersive component. The relative magnitudes involved depend on the geometry and the particular imposed conditions.

Blast Loading of a Rod

The physical parameters for this problem are motivated by the experimental work reported in Reference [181]. Let the rod be semi-infinite $(x \geq 0)$ in extent and experience a concentrated temperature blast at the free end $(x = 0)$. The boundary conditions for such a problem (again following Reference [15]) are

$$
\begin{aligned}
T(0,t) &= F(t) \\
\frac{\partial u(0,t)}{\partial x} &= -\alpha T(0,t) = -\alpha F(t)
\end{aligned}
\tag{3.28}
$$

In terms of the transforms these become

$$
\begin{aligned}
\hat{T}(0,\omega) &= \bar{A} + \bar{B} = 0 \\
\frac{d\hat{u}(0,\omega)}{dx} &= -ik_1 A - ik_2 B = -\alpha \hat{F}
\end{aligned}
\tag{3.29}
$$

where only that portion of the solution satisfying the radiation condition in $x > 0$ is considered. Solving for the coefficients and using equation(3.27) gives

$$\hat{u} = \frac{-i\alpha\hat{F}}{(k_1^2 - k_2^2)}\left\{k_1 e^{-ik_1 x} - k_2 e^{-ik_2 x}\right\}$$

$$\hat{T} = \frac{\hat{F}}{(k_1^2 - k_2^2)}\left\{\left[k_1^2 - \left(\frac{\omega}{c_o}\right)^2\right]e^{-ik_1 x} - \left[k_2^2 - \left(\frac{\omega}{c_o}\right)^2\right]e^{-ik_2 x}\right\} \qquad (3.30)$$

A quantity of interest is the strain, and this is obtained simply as

$$\hat{e} = \frac{d\hat{u}}{dx} = \frac{-\alpha\hat{F}}{(k_1^2 - k_2^2)}\{k_1^2 e^{-ik_1 x} - k_2^2 e^{-ik_2 x}\} = G(x, \omega)\hat{F} \qquad (3.31)$$

The strain histories can now be reconstructed using the inverse FFT, and the example plots of Reference [72] show the behavior as expected: the signal is propagating unattenuated at a constant speed. What might be surprising is its very small magnitude considering that the strain at the end is $\alpha T(0, t)$, which is on the order of 1000 $\mu\epsilon$. The explanation can be seen from the transfer function associated with the strain where it is noted that at $x = 0$, $G = \alpha$, but for non-zero x the second mode attenuates leaving

$$G(x, \omega) \approx -\frac{\alpha k_1^2}{k_1^2 - k_2^2}e^{-ik_1 x} \approx -\alpha\left(\frac{k_1}{k_2}\right)^2 e^{-ik_1 x}$$

That is, since k_2 is so much larger than k_1, the overall amplitude diminishes considerably.

Time reconstructions of the temperature histories at different locations show that beyond a very short distance from the source there is no temperature change. They also show that the propagating component of temperature could not be detected because the coefficient for the mode 1 is nearly zero.

Summary

This brief introduction to coupled thermoelasticity shows the versatility of the spectral analysis approach. It is capable of handling two quite distinct phenomena (i.e., stress and temperature) simultaneously under the same analytical framework. This is another example of the unity afforded by the approach.

3.6 Generalized Rod

It is very easy to generalize the descriptions of the rod without actually complicating the structure of the solution. In many cases it is only the spectrum relation that changes and otherwise the transfer functions are unchanged. Three types of modification are elaborated on: external influences, viscoelastic material behavior, and variable thickness.

External Influences

It can be imagined that as the wave moves down the rod, the response can be influenced by the surrounding medium. Specifically, let there be retarding forces proportional to velocity $(-\eta \dot{u})$ and displacement $(-Ku)$. The differential equation of motion becomes

$$EA\frac{\partial^2 u}{\partial x^2} - Ku - \eta\frac{\partial u}{\partial t} = \rho A\frac{\partial^2 u}{\partial t^2} \tag{3.32}$$

The spectral representation gives

$$EA\frac{d^2 \hat{u}}{dx^2} + [\omega^2 \rho A - K - i\omega\eta]\hat{u} = 0 \tag{3.33}$$

This has the same solution as equation (3.8) except that now the spectrum relation is complex and given by

$$k = \pm\left[\omega^2\frac{\rho A}{EA} - \frac{K}{EA} - i\omega\frac{\eta}{EA}\right]^{\frac{1}{2}} \tag{3.34}$$

The resulting waves, of course, are dispersive.

An interesting case arises when only the elastic constraint K is present, then there is the possibility of a cut-off frequency, i.e.,

$$\omega^2\frac{\rho A}{EA} - \frac{K}{EA} = 0 \quad \text{or} \quad \omega_o = \sqrt{\frac{K}{\rho A}} \tag{3.35}$$

Below this frequency the wavenumber is imaginary and therefore the wave attenuates. That is, for a spectrum wave the components below ω_o will not propagate and a signal sampled at a large distance will be rich in only the high-frequency components.

On the other hand, when only the velocity restraint is present, then the wavenumber is always complex indicating partial decay for all components. This is equivalent to damping. In the limit of small damping, the spectrum relation can be approximated as

$$k = \pm\sqrt{\frac{\rho A}{EA}}\left[\omega - i\frac{\eta}{2EA}\right]$$

Hence the generic solutions are dampened proportional to η and x. To add stability to all the calculations it is advisable to add some damping to the system. This is done in the sample programs in the Appendix by converting the wavenumber as

$$k = k_o(1 - i\eta), \qquad \eta \approx 0.005$$

where k_o would usually be the undamped value.

Reference [41] gives an interesting experimental account of the effect of coulomb friction on the outside wall of the rod.

Viscoelastic Rod

The purpose of this section is to show how waves in viscoelastic media may be posed in terms of spectral analysis. A good review of uniaxial wave propagation in viscoelastic rods (including some experimental results) can be found in Reference [127].

The derivation of the equation of motion follows the same procedure as before except that the stress/strain behavior is time dependent. While there are many ways of expressing this relationship (Reference [225], for example, uses hereditary integrals), the following form is adequate for present purposes:

$$\sum_p a_p \frac{d^p \sigma}{dt^p} = \sum_q b_q \frac{d^q e}{dt^q}, \qquad p, q = 0, 1, 2, \ldots \qquad (3.36)$$

Here the stress and strain are related through multiple derivatives in time. This can now be expressed in the spectral form as

$$\{\sum_p a_p (i\omega)^p\} \hat{\sigma} = \{\sum_q b_q (i\omega)^q\} \hat{e}$$

or simply

$$\hat{\sigma} = E(\omega)\hat{e}, \qquad E(\omega) \equiv \frac{\{\sum_q b_q (i\omega)^q\}}{\{\sum_p a_p (i\omega)^p\}} \qquad (3.37)$$

which resembles the linear elastic relation. The equation of motion also appears similar to the elastic case and is

$$E(\omega) A \frac{d^2 \hat{u}}{dx^2} + \omega^2 \rho A \hat{u} = -\hat{q} A$$

The only difference is that the spectrum relation is

$$k = \pm \omega \sqrt{\frac{\rho A}{E(\omega) A}} \qquad (3.38)$$

which is obviously dispersive. A plot of this (for the standard linear solid considered next) is shown in Figure 3.2. The non-zero imaginary component means there will be some attenuation and the effect of this is seen in the time reconstructions of Figure 3.10

To amplify on these results, consider, for concreteness, the special case of the standard linear solid as visualized by a parallel spring and dashpot in series with another spring and described by

$$\dot{\sigma} + \frac{\sigma}{\eta}[E_1 + E_2] = E_1 \dot{e} + \frac{E_1 E_2}{\eta} e$$

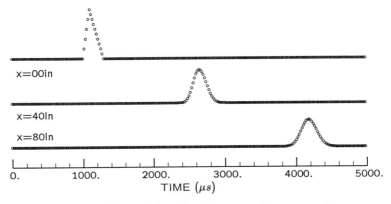

Figure 3.10: Effect of viscoelasticity on pulse propagation.

Unlike the linear elastic material, this requires three material properties E_1, E_2, η to describe it. The viscoelastic modulus is

$$E(\omega) = E_1 \frac{E_2 + i\omega\eta}{E_1 + E_2 + i\omega\eta} \tag{3.39}$$

and has the very slow and very fast behavior limits of

$$E(0) \approx \frac{E_1 E_2}{E_1 + E_2}, \qquad E(\infty) \approx E_1$$

Note that both of these limits are elastic. Consequently, the viscoelastic energy dissipation occurs only in the middle range. Reference [128] discusses this dissipation. The phase and group speeds in general are given as

$$c = \frac{\omega}{k} = \sqrt{\frac{EA}{\rho A}}, \qquad c_g = \frac{d\omega}{dk} = \sqrt{\frac{EA}{\rho A}} \bigg/ \left[1 - \frac{\omega}{2E}\frac{dE}{d\omega} \right] \tag{3.40}$$

and have the same limiting values. These speeds are shown plotted in Figure 3.3 where it is observed that the group speed has a maximum where there is significant damping. Thus damping does not necessarily case the wave to "slow down." The effect of viscosity on the propagation of a pulse is to decrease the amplitude (because of the damping) and spread the pulse out (because of the spectrum of speeds).

Rod of Variable Diameter

There is considerable interest in longitudinal wave propagation in members of varying cross-section. In References [121,124,136,139,167,175,199] are studies, both experimental and analytical, of waves in cones, hollow cones, shells, wedges and pyramids. While some of these have introduced new differential equations appropriate for the particular problem, it is believed that the following example captures the essence of the wave behavior when the cross-sectional properties change. Only the area will be assumed to change; some analysis of the more general case can be found in References [96,221]. When the diameter of the rod is not constant but varies slightly, then the spectral form of the differential equation is

$$E\frac{d}{dx}\left[A\frac{d\hat{u}}{dx}\right] + \omega^2 \rho A\hat{u} = 0 \tag{3.41}$$

This is a differential equation with variable coefficients but it cannot be integrated without assuming a specific form for the area variation $A(x)$.

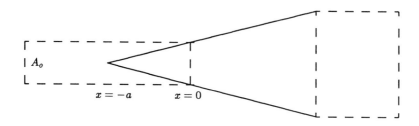

Figure 3.11: Geometry of variable area.

Assume the area varies according to the following power law

$$A(x) = A_o \left[\frac{a+x}{a}\right]^m \tag{3.42}$$

where A_o, a, m are constants as defined in Figure 3.11. Substituting into the differential equation and rearranging gives

$$\frac{d^2\hat{u}}{dx^2} + \left(\frac{m}{a+x}\right)\frac{d\hat{u}}{dx} + \beta^2\hat{u} = 0, \qquad \beta = \omega\sqrt{\frac{\rho A_o}{E A_o}} \tag{3.43}$$

This is a form of Bessel's equation and Appendix B shows that the complete solution is

$$\hat{u}(x) = \mathbf{A}z^\gamma J_\gamma(z) + \mathbf{B}z^\gamma Y_\gamma(z), \qquad \gamma = \tfrac{1}{2}(1-m), \qquad z = \beta(a+x) \tag{3.44}$$

Note that the spectrum relation is non-dispersive even though the signal itself is dispersive. That is, the disturbance travels at a constant speed but the amplitude changes. For the special case of $m = 1$, then,

$$\hat{u}(x) = \mathbf{A}J_o(z) + \mathbf{B}Y_o(z)$$

The asymptotic form of this result when $x \gg a$ is

$$
\begin{aligned}
\hat{u}(x) &= \mathbf{A}\sqrt{\frac{2}{\pi\beta x}}\cos\left(\beta x - \frac{\pi}{4}\right) + \mathbf{B}\sqrt{\frac{2}{\pi\beta x}}\sin\left(\beta x - \frac{\pi}{4}\right) \\
&= \frac{1}{\sqrt{\beta x}}\{\bar{\mathbf{A}}e^{-i\beta x} + \bar{\mathbf{B}}e^{i\beta x}\}
\end{aligned}
$$

which is similar in form to that for the uniform rod except for the decreasing amplitude. On the other hand, for the cone (or linear taper) $m = 2$ and noting the relation between the Bessel functions of half order and the trigonometric functions gives

$$
\hat{u}(x) = \frac{1}{\sqrt{\beta(a+x)}}\{\bar{\mathbf{A}}e^{-i\beta(a+x)} + \bar{\mathbf{B}}e^{i\beta(a+x)}\}
$$

which is the same as above except that it is not restricted to the asymptotic form. Note that the special case of the uniform rod ($m = 0$) can also be recovered by utilizing the properties of the Bessel functions of half order.

Of course, once the generic solution is in hand, then the transfer function can be obtained just as in the previous cases. It is then a simple matter to obtain the time reconstructions by performing the inverse FFT.

3.7 Mindlin-Herrmann Rod Theory

The elementary rod theory considers only the longitudinal motion, but it is known that due to the Poisson's ratio effect there is some lateral motion also. Obviously some energy goes into moving these particles laterally and therefore it must affect the wave behavior. The incorporation of this lateral effect is the improvement in the rod theory considered here.

There are many ways of deriving higher-order structural theories. The one chosen here and later for beams and plates is first to derive the equations of motion directly in terms of resultants. Then an assumption is made concerning the deformation field in such a way as to add degrees of freedom. It is these degrees of freedom that add more modes in the response. This approach has the advantage of being a direct extension of the elementary theory and therefore, at least, these elementary results are recoverable. The underlying assumption is that "higher-order effects" are only perturbations on the elementary solutions — this, of course, may not always be true.

Equation of Motion

The equation for the axial motion is the same as in the elementary case. That is,

$$
\frac{\partial F}{\partial x} = \rho A \ddot{u}
$$

To describe the lateral motion, consider the equations of motion of the pie-shaped element of Figure 3.12 under the resultant normal force per unit length S and shear force per unit wedge angle Q. Summing forces in the radial direction gives

$$-Q\Delta\theta + (Q + \Delta Q)\Delta\theta - S\Delta x\Delta\theta = \rho\tfrac{1}{2}a^2\Delta\theta\Delta x\, \ddot{u}_{rc}$$

or

$$\frac{\partial Q}{\partial x} - S = \tfrac{1}{2}\rho a^2\, \ddot{u}_{rc} \tag{3.45}$$

It is assumed here that the cross-section is nearly circular (of average radius a) so that any shear on the side faces can be neglected. Also, u_{rc} is the displacement in the radial direction of the centroid of the wedge. There are now two equations of motion in contrast to the single equation of the elementary theory.

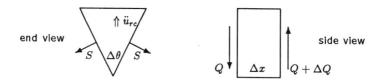

Figure 3.12: Resultants acting on rod segment.

The simplest deformation field consistent with the axial motion is to assume

$$u = u(x, t), \qquad u_r = r\phi(x, t), \qquad u_\theta = 0 \tag{3.46}$$

where r is a radial distance from some central position and ϕ is the extra degree of freedom. Thus, the lateral displacements are zero at the center and a maximum on the outside. The stresses of interest can be obtained using Hooke's law to give

$$\sigma_x = 2\lambda\phi + (\lambda + 2\mu)\frac{\partial u}{\partial x}$$

$$\sigma_\theta = 2(\lambda + \mu)\phi + 2\mu\frac{\partial u}{\partial x}$$

$$\sigma_{r\theta} = \mu r\frac{\partial \phi}{\partial x}$$

where λ, μ are the Lamé constants — more elaboration on these can be found in Chapter 6. Note that only the shear stress is dependent on r. These, of course, are unrealistic distributions, but they are adequate for the present purpose since they will be used only to obtain resultants. These resultants are, in fact,

$$F = \int_A \sigma_z\, dA = \left[2\lambda\phi + (\lambda + 2\mu)\frac{\partial u}{\partial x}\right]A$$

$$S = \int \sigma_\theta dr = K_1^2 \left[2(\lambda + \mu)\phi + \lambda \frac{\partial u}{\partial x} \right] a$$

$$Q = \int \sigma_{r\theta} r dr = K^2 \left[\mu \frac{\partial \phi}{\partial x} \right] \frac{a^3}{3} \tag{3.47}$$

The constants K_1 and K are correction factors introduced to account both for the nonsymmetry of the cross-section and the fact that the actual stresses are not distributed as assumed above. Obtaining numerical values for them is discussed later but they are of order unity.

The Mindlin-Herrmann equations for the rod can now be obtained by substituting for the resultants into the equations of motion to get

$$(\lambda + 2\mu)\frac{\partial^2 u}{\partial x^2} + 2\lambda \frac{\partial \phi}{\partial x} = \rho \ddot{u}$$

$$\tfrac{4}{3}K^2 \mu a^2 \frac{\partial^2 \phi}{\partial x^2} - 8K_1^2(\lambda + \mu)\phi - 4K_1^2\lambda \frac{\partial u}{\partial x} = \rho a^2 \ddot{\phi} \tag{3.48}$$

These are a set of coupled equations in the longitudinal displacement $u(x,t)$ and lateral contraction $\phi(x,t)$, and were first presented in Reference [162]. The elementary theory is recovered by setting the resultant S to zero and replacing ϕ in terms of u in the first of the above equations.

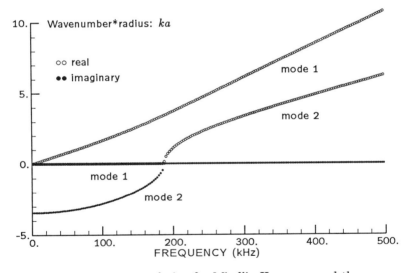

Figure 3.13: Spectrum relation for Mindlin-Herrmann rod theory.

Spectrum Relation

Since there are two dependent variables u and ϕ and the coefficients are constant, then assume solutions of the form

$$u = u_o e^{-i(kz - \omega t)}, \qquad \phi = \phi_o e^{-i(kz - \omega t)}$$

On substitution into the differential equations these give the system to be satisfied as

$$\begin{bmatrix} -(\lambda + 2\mu)k^2 + \rho\omega^2 & -i2\lambda k \\ i4K_1^2\lambda k & -\tfrac{4}{3}K^2\mu a^2 k^2 - 8K_1^2(\lambda + \mu) + \rho a^2 \omega^2 \end{bmatrix} \begin{Bmatrix} u_o \\ \phi_o \end{Bmatrix} = 0 \quad (3.49)$$

As was done in the coupled thermoelastic problem, setting the determinant to zero gives the characteristic equation for determining k as

$$k^4 \left[\tfrac{4}{3}K^2(ak_o\gamma)^2 \right] + k^2 k_p^2 \left[8K_1^2 - \tfrac{4}{3}K^2(ak_o\gamma)^2 \right] - k_p^2 k_o^2 \left[8K_1^2 - (ak_s\gamma)^2 \right] = 0 \quad (3.50)$$

where each term in square brackets is non-dimensional and the following wavenumbers were introduced:

$$k_o = \omega/c_o, \quad k_p = \omega/c_p, \quad k_s = \omega/c_s$$

associated with the speeds

$$c_p = \sqrt{\frac{2\mu + \lambda}{\rho}}, \quad c_s = \sqrt{\frac{\mu}{\rho}}, \quad \gamma \equiv \sqrt{\frac{c_s}{c_p}}$$

The notation for the speeds is presented in Chapter 6 but for now they can be treated just as constants. The characteristic equation is quadratic in k^2 and therefore, in all, there are four modes in contrast to the two of the elementary solution. Solving in the usual way for a quadratic equation, it is possible to obtain the spectrum relations, and Figures 3.13 and 3.14 show the typical spectrum and dispersion behaviors.

The mode 1 behavior is real only, but it shows a decreasing speed with frequency. It is interesting to note how the group speed has a minimum and then asymptotes to the phase speed. In the limit of low frequencies, this mode reduces to that of the elementary theory. The second mode behavior is quite different. First, it exhibits a cut-off frequency given by

$$w_o = \sqrt{8c_p c_s} \, K_1 \frac{1}{a}$$

Below this frequency the second mode attenuates. Where this occurs is also inversely proportional to the radius a. Thus slender rods give a high w_o and so the second propagating component of the wave is usually not observed. Further, even though the mode is present at the low frequencies, the large value of k_2 means that it is highly attenuated for any non-zero x; consequently, it is usual to neglect it.

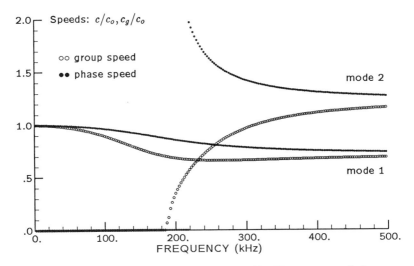

Figure 3.14: Dispersion relation for the Mindlin-Herrmann rod theory.

Solution Procedures

Since there are four modes in the solution and it is a coupled system, then, as in thermoelasticity, the solution can be written as

$$\hat{u}_o = \mathbf{A}e^{-ik_1z} + \mathbf{B}e^{-ik_2z} + \mathbf{C}e^{ik_1z} + \mathbf{D}e^{ik_2z} \tag{3.51}$$
$$\hat{\phi}_o = \bar{\mathbf{A}}e^{-ik_1z} + \bar{\mathbf{B}}e^{-ik_2z} + \bar{\mathbf{C}}e^{ik_1z} + \bar{\mathbf{D}}e^{ik_2z}$$

It is sufficient to concentrate on either the axial deformation or the radial deformation and use the amplitude ratio given by

$$\frac{\phi_o}{u_o} = \frac{-(\lambda + 2\mu)k^2 + \rho\omega^2}{i2\lambda k} \tag{3.52}$$

to obtain the other quantity.

The boundary conditions are specified in terms of the resultants F and S and the displacements u and ϕ. For example, axial impact could be specified as

$$F = \left[2\lambda\phi + (\lambda + 2\mu)\frac{\partial u}{\partial x}\right]A = F(t)$$

$$S = K_1^2\left[2(\lambda + \mu)\phi + \lambda\frac{\partial u}{\partial x}\right]a = 0$$

and then by substituting for the displacements, the coefficients $\mathbf{A}, \mathbf{B}, \bar{\mathbf{A}}, \bar{\mathbf{B}}$ (since there are only forward moving waves) can be solved for. The procedure is similar to that for the coupled thermoelastic problem. It is also similar to the procedures necessary for solving beam problems in the next chapter, and so explicit examples will be left until then.

Determining K_1 and K

Consider the limiting cases of low and high frequency (or what is the same thing, small and large radius a, respectively). That is, for low frequencies

$$k = k_o \, , \, \frac{i}{a} \quad \text{or} \quad c = c_o \, , \, 0$$

The first in each case is the elementary solution. Notice that for the second mode, the wavenumber depends on the reciprocal of the radius (the actual relation is more complicated, but this is a fair approximation for the present discussion). For high frequencies

$$k = \frac{k_s}{\sqrt{\frac{4}{3}K}} \, , \, k_p \quad \text{or} \quad c = c_s \sqrt{\frac{4}{3}K} \, , \, c_p$$

The exact solution for high frequencies as determined from the Pochhammer-Chree theory (Reference [142]) gives the limiting speeds as

$$c = c_R \quad \text{and} \quad c = c_p \tag{3.53}$$

where these are the Rayleigh and compressional wave speeds, respectively. (More details can be found in Chapter 6.) Therefore, K can be chosen so that the first of these limits is satisfied, that is, $K = 0.802$. The other coefficient does not enter either limit and so is usually taken as unity. References [236] considers this discussion in much more detail.

Comparisons with Other Approximate Theories

There are many specialized rod theories. As an example, the Love theory for rods (as used in Reference [50]) also takes lateral inertia effects into account and gives for the axial motion

$$c_o^2 \frac{\partial^2 u}{\partial x^2} - \frac{\partial^2 u}{\partial t^2} + \frac{\nu a^2}{2} \frac{\partial^4 u}{\partial x^2 \partial t^2} = 0 \tag{3.54}$$

where a is the rod radius and ν is the Poisson's ratio. This recovers the simple wave equation when $\nu = 0$ or $a = 0$. The second condition is equivalent to saying that the radius should be small compared to the wavelengths extent. The spectrum and dispersion relations for this equation are

$$k = \frac{\omega}{\sqrt{c_o^2 - \frac{1}{2}(\nu a \omega)^2}} \, , \qquad c = \sqrt{c_o^2 - \frac{1}{2}(\nu a \omega)^2} \tag{3.55}$$

showing that it is possible to have zero phase velocity. This approximation is poor in comparison to the Mindlin-Herrmann theory but it has been used because of its apparent simplicity. Notice also that it is a single mode solution.

Chapter 4

Flexural Waves in Beams

Besides the rod, the other major structural component of importance is the beam. The purpose of this component is to support lateral loads and, consequently, the displacement is off the centerline. The dynamic behavior generated is called *flexural motion*.

From the wave point of view the significant difference between a beam and a rod is that the beam (even in its simplest form) does not have a D'Alembert solution and therefore the solutions are dispersive from the outset. Also (because of the higher order governing equations), there are two fundamental modes of motion, one of which propagates while the other is evanescent.

The development in this chapter is similar to that of the last. That is, after deriving the basic solution, its application to reflections, tapered beams, and so on is demonstrated. Since the analytical framework is the same as for rods, these applications are not described in as much detail as before. Instead, the chapter concentrates on two new topics. The first is that of curved beams and the other is remote sensing. Both of these bring out new aspects of spectral analysis, namely, numerical evaluation of the spectrum relation and explicit use of the phase.

Only the source code for the program ROOTCOMP (used to obtain the roots of a characteristic equation) is given in Appendix D — the source code for the other programs including BEAMCOMP is available through the author.

4.1 Bernoulli-Euler Beam Theory

Consider a long, narrow beam with the loads applied as in Figure 4.1. The Bernoulli-Euler beam model assumes that only bending moment $M(x,t)$ and shear force $V(x,t)$ resultants act at each section. It also assumes that the deflection of the centerline $v(x,t)$ is small and only transverse. While this theory assumes the presence of a transverse shear force, it neglects any deformation due to it. This is rectified in the higher-order Timoshenko beam theory introduced later in the chapter. An alternative development for both theories is given in Reference [190].

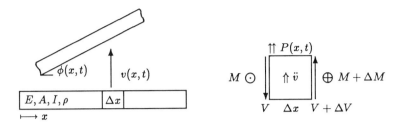

Figure 4.1: Narrow beam in flexure and typical loaded element.

Equation of Motion

The equations of motion in the y-direction and about the z-axis are (if Δx is small), respectively,

$$\frac{\partial V}{\partial x} = \rho A \frac{\partial^2 v}{\partial t^2} - P \tag{4.1}$$

$$\frac{\partial M}{\partial x} + V = \rho I_G \frac{\partial^2 \phi}{\partial t^2} \tag{4.2}$$

If the material is linear elastic, then $\sigma = Ee$, and integrating on the cross-section to get the resultant moment gives the moment-deflection relation as

$$M = EI \frac{\partial^2 v}{\partial x^2}$$

Combining this with the equations of motion and neglecting the rotational inertia term $\rho I \ddot{\phi}$, gives

$$\frac{\partial^2}{\partial x^2} \left[EI \frac{\partial^2 v}{\partial x^2} \right] + \rho A \frac{\partial^2 v}{\partial t^2} = P \tag{4.3}$$

It can be easily verified that there is no D'Alembert solution to the homogeneous part of this equation even if the cross-sectional properties are constant.

Spectral Analysis

Consider the beam to have constant properties along its length, then the homogeneous differential equation can be written in the spectral form

$$\frac{d^4 \hat{v}}{dx^4} - \beta^4 \hat{v} = 0, \qquad \beta^2 = \omega \sqrt{\frac{\rho A}{EI}} \tag{4.4}$$

Particular solutions to this can be obtained from particular solutions of the following two equations:

$$\frac{d^2\hat{v}}{dx^2} - \beta^2\hat{v} = 0, \qquad \frac{d^2\hat{v}}{dx^2} + \beta^2\hat{v} = 0 \qquad (4.5)$$

This form emphasizes that the beam has two fundamentally different modes — this will become apparent later.

Since the coefficients are constants, then the solutions are obviously of the exponential form $e^{\pm ikx}$, which on substituting into the equations gives

$$k_1 = \pm\beta, \qquad k_2 = \pm i\beta$$

There are four possibilities in all, allowing the complete solution to be written as

$$v(x,t) = \sum\{Ae^{-ikx} + Be^{-kx} + Ce^{ikx} + De^{kx}\}e^{i\omega t}, \qquad k = \sqrt{\omega}\left[\frac{\rho A}{EI}\right]^{1/4} \qquad (4.6)$$

The first and third are wave solutions while the second and fourth are spatially damped vibrations (sometimes called the ringing terms).

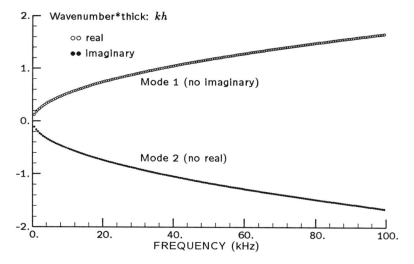

Figure 4.2: Spectrum relation for beams.

The spectrum relations for both modes are plotted in Figure 4.2. Since the mode 2 behavior is entirely imaginary, then there is no propagation behavior for this mode. Considering only the mode 1 wave motion gives the phase and group speeds as

$$c \equiv \frac{\omega}{k} = \sqrt{\frac{EI}{\rho A}}k = \sqrt{\omega}\left[\frac{EI}{\rho A}\right]^{1/4}$$

$$c_g \;\equiv\; \frac{d\omega}{dk} = 2\sqrt{\frac{EI}{\rho A}}\,k = 2\sqrt{\omega}\left[\frac{EI}{\rho A}\right]^{1/4} = 2c \qquad (4.7)$$

Both of these are shown plotted in Figure 4.3. In contrast to the elementary rod case, these speeds depend on the sectional properties of the beam in terms of the area $A = bh$ and second moment of area $I = bh^3/12$, where b is the width and h the depth. Note that the group speed is twice that of the phase. Both speeds are also dispersive and indicate speeds approaching infinity for very high frequencies. This unreasonable limit will be corrected later in this chapter by use of the Timoshenko beam theory.

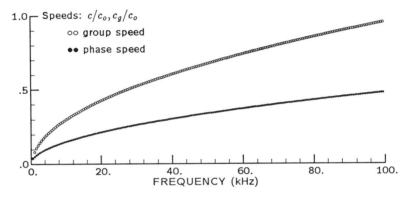

Figure 4.3: Dispersion relations for beams.

4.2   Basic Solution for Waves in Beams

Since the spectrum relation is dispersive, then (as pointed out before) there is no D'Alembert solution and, consequently, it is difficult to describe the wave in the time domain. This is where the spectral approach is most beneficial because the frequency domain is the most natural setting in which to discuss the properties of dispersive systems.

Stress and Shear Behavior

Using the subscripts i and r to represent the forward- and backward-moving waves respectively, then the two groups can be expressed as

$$v_i = \sum \mathbf{A}e^{-i(kx-\omega t)} + \sum \mathbf{B}e^{-kx-i\omega t}, \qquad v_r = \sum \mathbf{C}e^{i(kx+\omega t)} + \sum \mathbf{D}e^{kx+i\omega t}$$

To aid in the analysis, assume that the wave originates at $x = 0$ and propagates symmetrically outward, then

$$\text{at } x = 0: \qquad \frac{\partial v}{\partial x} = 0, \qquad v^- = v^+$$

giving, simply, $\mathbf{B} = -i\mathbf{A}$, $\mathbf{D} = -i\mathbf{C}$ and $\mathbf{C} = \mathbf{A}$. Consequently, the two wave groups are written in terms of a single amplitude spectrum as

$$v_i = \sum \mathbf{A}[e^{-ikx} - ie^{-kx}]e^{i\omega t}, \qquad v_r = \sum \mathbf{A}[e^{ikx} - ie^{kx}]e^{i\omega t} \qquad (4.8)$$

The terms in square brackets are the transfer functions for the waves and they indicate a changing amplitude with respect to position. For the forward disturbance, the other mechanical quantities of usual interest are

Strain: $e_i = \sum \dfrac{h}{2}k^2 A[e^{-ikx} + ie^{-kx}]e^{i\omega t}$

Shear: $V_i = \sum iEIk^3 A[e^{-ikx} + e^{-kx}]e^{i\omega t}$

Velocity: $\dot{v}_i = \sum i\omega A[e^{-ikx} - ie^{-kx}]e^{i\omega t} = \sum ik^2 \sqrt{\dfrac{EI}{\rho A}} A[e^{-ikx} - ie^{-kx}]e^{i\omega t}$

It is apparent that, in general, there is no simple interrelationship among these quantities. However, at the origin of the wave $(x = 0)$

$$e_i = \frac{h}{2}[1 + i]\sum k^2 \mathbf{A}, \qquad \dot{v}_i = \sqrt{\frac{EI}{\rho A}}[i + 1]\sum k^2 \mathbf{A}$$

showing that always

$$\dot{v}_i(o, t) = \frac{2}{h}\sqrt{\frac{EI}{\rho A}}e_i(o, t) = \frac{2}{Eh}\sqrt{\frac{EI}{\rho A}}\sigma_i(o, t) = \frac{c_o}{\sqrt{3}}e_i(o, t) \qquad (4.9)$$

That is, the particle velocity has a simple relationship to the bending strain and stress. This, of course, is not true at any other position.

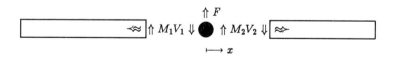

Figure 4.4: Free body diagrams for impact of a beam.

Force History Applied to Center of Beam

Consider a very long beam impacted at its center. Let the center of the beam be at $x = 0$ and the force be applied to an imaginary rigid, massless, joint as shown in Figure 4.4. The conditions at this joint require

$$\frac{\partial v}{\partial x} = 0$$
$$-F + 2V = m_J \ddot{v}_J = 0 \tag{4.10}$$

Substituting for these quantities in terms of the displacement solution gives the outward wave, for example, as

$$v(x,t) = \sum \frac{i\hat{F}}{4EIk^3}[e^{-ikz} - ie^{-kz}]e^{i\omega t} \tag{4.11}$$

The behavior of the strain at various positions along the beam is shown in Figure 4.5. The response is obviously dispersive. Note particularly that because the low-frequency components travel at nearly zero speed, they will take the whole window to arrive at the monitoring site. As a result, an extra-large window is usually required when analyzing beams. Indeed, the window used for the present analysis is twice that shown. Since this is a recurring problem with beams, it is of value to consider it in more detail.

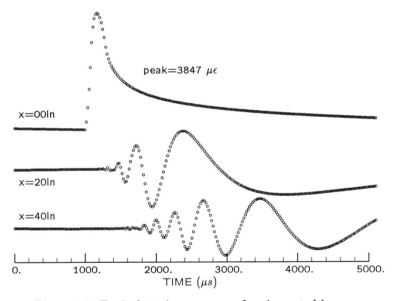

Figure 4.5: Typical strain responses of an impacted beam.

The transfer function for the wave is

$$G(x,\omega) = \frac{i}{4EIk^3}[e^{-ikx} - ie^{-kx}]$$

and this is different from the rod (see section (3.2)) in two significant respects. First, the presence of e^{-kx} indicates a large contribution at $x = 0$ that dies rapidly as x increases. Secondly, and perhaps more important, the transfer function depends on the reciprocal of k^3 and consequently $\omega^{3/2}$ also. This fractional power dependence on frequency does not really pose a problem in the spectral analysis approach *per se*, although it is what makes the analysis in the time domain difficult. For example, the relation between the force and strain at the impact site is

$$\hat{e}(0,\omega) = \frac{h}{8EI}(1-i)\frac{\hat{F}}{k} = \frac{h}{8EI}\left[\frac{EI}{\rho A}\right]^{1/4}(1-i)\frac{\hat{F}}{\sqrt{\omega}} \qquad (4.12)$$

The convolution theorem of the first chapter (equation (1.12)) can be used to convert this to time domain form as

$$e(0,t) = \frac{h}{4EI}\left[\frac{EI}{\rho A}\right]^{1/4}\frac{1}{\sqrt{2\pi}}\int_o^t \frac{F(\tau)}{\sqrt{t-\tau}}d\tau = A\int_o^t \frac{F(\tau)}{\sqrt{t-\tau}}d\tau \qquad (4.13)$$

This interesting equation (which also appears in Reference [201]) shows the strain (and velocity) to be a singular hereditary integral of the force. Consequently, the response will not be quiescent until the time approaches infinity. This is where the fractional powers cause some windowing problems — no matter what size the window is, some response will always persist beyond it and thus move into the neighboring window. The negative effects of this can be minimized by choosing a relatively large window size but this is at the cost of more computation time. A cheaper approach is to include some damping in the system so that all signals eventually die out. Actually, a judicious combination of both will give the best results.

4.3 Boundary Reflections of Flexural Waves

The examples in this section reinforce the contention that the same spectral analysis methodology can be used irrespective of the particular structural model used. Results parallel to those of the rod are developed for typical boundary conditions. What must be realized, however, is that since there are more modes, then there is also a wider variety of boundary conditions.

Boundary Conditions

Let the end of a semi-infinite beam be at $x = 0$ and let the incident wave be from left to right. A complete harmonic solution to the beam equation can be written in the form

$$\hat{v}(x,\omega) = \mathbf{A}e^{-ikz} + \mathbf{B}e^{-kz} + \mathbf{C}e^{ikz} + \mathbf{D}e^{kz}, \qquad k = k(\omega)$$

The first two terms are the incident wave and $\mathbf{A}, \mathbf{B}$ and ω are assumed specified. The remaining terms are the generated reflections. The last term $(\mathbf{D}e^{kz})$, while not a wave, is nonetheless necessary to satisfy the boundary conditions even if only a propagating term is incident on the boundary. As emphasized before, the above is the general solution to the governing differential equation and as such there cannot be any others. That is, these are the totality of possible responses for the given structural model and properly stated boundary conditions should then determine the appropriateness of each.

The boundary conditions are specified in terms of the different orders of derivative of $\hat{v}(x,\omega)$. These are collected as

$$
\begin{aligned}
\text{Displacement:} &\quad \hat{v} = \hat{v}(x,\omega) \\
\text{Slope:} &\quad \hat{\phi} = \frac{d\hat{v}}{dx} \\
\text{Moment:} &\quad \hat{M} = EI\frac{d^2\hat{v}}{dx^2} \\
\text{Shear Force:} &\quad \hat{V} = -EI\frac{d^3\hat{v}}{dx^3}
\end{aligned}
$$

Table 4.1 is a collection of some typical simple boundaries and the associated equations. Note that in each case two conditions must be specified in the time domain and these become conditions on the various space derivatives in the frequency domain.

Pinned End

The simplest boundary condition is that of a pinned end. Here the deflection is zero giving as one of the equations

$$\mathbf{A} + \mathbf{B} + \mathbf{C} + \mathbf{D} = 0$$

There is also no moment giving as the second equation

$$EI[\mathbf{A}(-ik)^2 + \mathbf{B}(-k)^2 + \mathbf{C}(ik)^2 + \mathbf{D}(k)^2] = 0$$

These two equations allow $\mathbf{C}$ and $\mathbf{D}$ to be solved for giving (after simplifying and canceling the wavenumber k)

$$\mathbf{C} = -\mathbf{A}, \qquad \mathbf{D} = -\mathbf{B} \tag{4.14}$$

Thus, the reflected wave is the same as the incident but inverted. This is the same as occurs for the rod. Note that there is no spatially decaying term generated if none is incident.

BC Type	Imposed	conditions
Fixed	$v(0,t) = 0$,	$\dfrac{\partial v(0,t)}{\partial x} = 0$
Pinned	$v(0,t) = 0$,	$\dfrac{\partial^2 v(0,t)}{\partial x^2} = 0$
Free	$EI\dfrac{\partial^2 v(0,t)}{\partial x^2} = 0$,	$EI\dfrac{\partial^3 v(0,t)}{\partial x^3} = 0$
Linear Spring	$EI\dfrac{\partial^2 v(0,t)}{\partial x^2} = 0$,	$EI\dfrac{\partial^3 v(0,t)}{\partial x^3} = -K\,v(0,t)$
Torsion Spring	$v(0,t) = 0$,	$EI\dfrac{\partial^3 v(0,t)}{\partial x^3} = -\alpha\,\dfrac{\partial v(0,t)}{\partial x}$
Dashpot	$EI\dfrac{\partial^2 v(0,t)}{\partial x^2} = 0$,	$EI\dfrac{\partial^3 v(0,t)}{\partial x^3} = -\eta\,\dfrac{\partial v(0,t)}{\partial t}$
Mass	$EI\dfrac{\partial^2 v(0,t)}{\partial x^2} = 0$,	$EI\dfrac{\partial^3 v(0,t)}{\partial x^3} = -m\,\dfrac{\partial^2 v(0,t)}{\partial t^2}$

Table 4.1: Some typical boundary conditions for beams.

Free End

An example of a more complicated boundary is that of a free end. It is necessary to impose both no moment, as above, and also no shear, giving

$$EI[\mathbf{A}(-ik)^2 + \mathbf{B}(-k)^2 + \mathbf{C}(ik)^2 + \mathbf{D}(k)^2] = 0$$
$$EI[\mathbf{A}(-ik)^3 + \mathbf{B}(-k)^3 + \mathbf{C}(ik)^3 + \mathbf{D}(k)^3] = 0$$

Again k can be canceled through allowing $\mathbf{C}$ and $\mathbf{D}$ to be simply solved for as

$$\mathbf{C} = -i\mathbf{A} + (1+i)\mathbf{B}, \qquad \mathbf{D} = (1-i)\mathbf{A} + i\mathbf{B} \qquad (4.15)$$

Note that even if only a traveling wave is incident (i.e., $\mathbf{B} = 0$), both a traveling wave ($\mathbf{C}$) and boundary vibration ($\mathbf{D}$) are generated. The presence of the complex i means there is also a phase shift since $-i = e^{-i\pi/2}$. However, the responses do not depend on frequency, hence there is no additional dispersion.

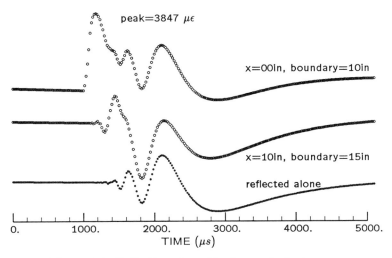

Figure 4.6: Reflections from the free end of a beam.

Figure 4.6 shows an example of the response near a free end. The impact site $(x = 0$ in.$)$ and monitoring site (x) were chosen such that the traveled path in both cases is the same, i.e., 20 inches. As a result the reflections are the same for both cases — only the initial portions are different. Parenthetically, a side benefit of this approach to waves is that the decomposition of the response is a natural by product. That is, the reflected wave can be reconstructed on its own, as shown, if required.

Reflection from an Elastic Boundary

Assume that at the end of the beam there are two elastic springs, one resisting the deflection and the other resisting the rotation. Let the incident wave contain a ringing term also, then shear and moment balance equations require

$$EI\frac{d^3\hat{v}}{dx^3} = -K\hat{v}\,, \qquad EI\frac{d^2\hat{v}}{dx^2} = -\alpha\frac{d\hat{v}}{dx}$$

giving, with $K^* \equiv k^3 K/EI$, $\alpha^* \equiv k\alpha/EI$,

$$\begin{bmatrix} (i - K^*) & -(1 + K^*) \\ -(1 - i\alpha^*) & (1 + \alpha^*) \end{bmatrix} \begin{Bmatrix} C \\ D \end{Bmatrix} = \begin{bmatrix} (i + K^*) & -(1 - K^*) \\ (i + i\alpha^*) & -(l - \alpha^*) \end{bmatrix} \begin{Bmatrix} A \\ B \end{Bmatrix} \tag{4.16}$$

Since K^* and α^* depend on k (and hence ω), then this relation is frequency dependent. As a consequence, the response will exhibit the same type of dispersion effect as observed in the rod case. That is, for given springs, the low frequencies experience a fixed end, whereas the high frequencies experience a free end. The special cases of free end $(K = 0, \alpha = 0)$ and fixed end $(K = \infty, \alpha = \infty)$ are easily recovered from the above.

4.4   Reflections and Transmissions

The discontinuities of particular interest here are the massy joint, the collinear stepped beam, and the angled joint. In each of these cases, let the flexural displacements for beam 1 and 2 be

$$
\begin{aligned}
\hat{v}_1(x) &= \mathbf{A}e^{-ik_1 x} + \mathbf{C}e^{ik_1 x} + \mathbf{D}e^{k_1 x} \\
\hat{v}_2(x) &= \bar{\mathbf{A}}e^{-ik_2 x} + \bar{\mathbf{B}}e^{-k_2 x}
\end{aligned}
\tag{4.17}
$$

where it is assumed that only a propagating wave is incident on the joint and its amplitude spectrum is represented by the $\mathbf{A}$ term. This is reasonable since in most cases the generated evanescent term decays to negligible size. Also retaining this term is cumbersome. In fact, it does not take much to make these types of problems unwieldy and that is why the following chapter develops a matrix method approach to handling multi-member problems. In this chapter, however, the simpler versions are treated so as to be able to emphasize the essential mechanics of the problem. The first beam also contains a reflected wave $\mathbf{C}$ and a ringing term $\mathbf{D}$, and the second beam contains corresponding propagating and ringing terms. As pointed out before, this is the maximum number of possibilities — the boundary conditions should determine if any are redundant. The angled joint introduces the idea of mode conversion; for example, an incident longitudinal wave generating both longitudinal and flexural waves. Additional terms must then be appended to the above.

The companion problem of waves traveling in periodic structures is treated in References [97,144,151]. Some additional considerations of waves traveling in connected structures are given in Chapter 5.

Concentrated Mass

Consider two beams of similar sectional properties connected by a rigid concentrated mass as shown in Figure 4.4. Continuity and balance of resultants at the concentrated mass requires

$$
\begin{aligned}
v_1 = v_2 &= v_J \\
\frac{\partial v_1}{\partial x} = \frac{\partial v_2}{\partial x} &= \phi_J \\
-M_1 + M_2 &= 0 \\
-V_1 + V_2 &= m_J \ddot{v}_J
\end{aligned}
\tag{4.18}
$$

where the subscript J refers to the joint properties. This gives with $m^* \equiv m_J/EIk^3$ and assuming both beam segments to be the same

$$
\begin{bmatrix}
1 & 1 & -1 & -1 \\
i & 1 & i & 1 \\
-1 & 1 & 1 & -1 \\
-i & 1 & i+m^*\omega^2 & i+m^*\omega^2
\end{bmatrix}
\begin{Bmatrix}
C \\ D \\ \bar{A} \\ \bar{B}
\end{Bmatrix}
=
\begin{Bmatrix}
-1 \\ i \\ 1 \\ 0
\end{Bmatrix} A
\tag{4.19}
$$

The presence of the mass makes the equation frequency dependent and thus the mass (as in the case of rods) acts as a frequency filter. Low-frequency components transmitt through the joint unaffected, the high-frequency components are blocked.

Collinear Beams

The analysis will be restricted to beams that have discontinuity of stiffness and density, but remain collinear. This is the analog of the stepped rod, and the free body diagram is similar to that shown in Figure 4.4. The joint is assumed to be rigid, of very small mass and extent. A fuller exposition can be found in Reference [58].

Continuity conditions at the joint are the same as for the concentrated mass except that the joint mass is considered zero. In terms of the displacement coefficients these conditions become (with $x = 0$ being the location of the joint)

$$
\begin{bmatrix}
1 & 1 & -1 & 1 \\
iR & R & iQ & Q \\
-1 & 1 & R^2Q^2 & -R^2Q^2 \\
-i & 1 & -iRQ^3 & RQ^3
\end{bmatrix}
\begin{Bmatrix}
C \\ D \\ \bar{A} \\ \bar{B}
\end{Bmatrix}
=
\begin{Bmatrix}
-1 \\ iR \\ 1 \\ -i
\end{Bmatrix} A
\tag{4.20}
$$

where the coefficient ratios $R \equiv [E_2I_2/E_1I_1]^{1/4}$ and $Q \equiv [\rho_2A_2/\rho_1A_1]^{1/4}$ have been introduced. Generally, it is best to solve this system numerically as part of the general scheme for obtaining the transfer function. The explicit solution is given here because it can be used to help verify the implicit solution schemes used on some of the more general cases. Solving in terms of A gives

$$
C = [2RQ(R^2 - Q^2) - i(1 - R^2Q^2)^2]\frac{1}{D}A
$$

$$
D = (1 - R^2Q^2)(1 + R^2Q^2)[1 - i]\frac{1}{D}A
$$

$$
\bar{A} = 2(1 + R^2Q^2)(R + Q)\frac{1}{DQ}A
$$

$$
\bar{B} = -2(1 - R^2Q^2)(R + iQ)\frac{1}{DQ}A
$$

$$
D = 2RQ(R^2 + Q^2) + (1 + R^2Q^2)^2
\tag{4.21}
$$

Without actually making plots, it is possible to conclude some interesting features about the response. First, it is noticed from the transfer functions for $\mathbf{C}$ and $\bar{\mathbf{A}}$ that while the reflected wave may have a phase change, the transmitted wave is always in phase with the incident wave (since the transmission coefficient for the latter is always real). These expressions do not contain the frequency explicitly and therefore the step does not act as a frequency filter. Again this result parallels the case of the rod. The limiting cases of a free end $(R = 0)$ and a fixed end $(R = \infty)$ allow the previous results to be recovered. What is interesting is the interpretation of the transmitted waves. For example, for the free end

$$\bar{\mathbf{A}} = 2\mathbf{A}, \qquad \bar{\mathbf{B}} = -2i\mathbf{A}$$

The displacement "overshoots" and gives twice the incident value. This displacement amplification is worth keeping in mind when considering the response of connected systems.

The effects of an elastic joint can be analyzed similarly to above. On the incident side there are $\mathbf{A}, \mathbf{C}, \mathbf{D}$ terms and on the transmitted side $\bar{\mathbf{A}}$ and $\bar{\mathbf{B}}$ terms. The joint, however, must contain all four $\mathbf{A}, \mathbf{B}, \mathbf{C}, \mathbf{D}$ terms thus indicating the generation of a standing wave. The results are analogous to those of the elastic joint in a rod but it is difficult to detect the multiple reflections in the time domain since all signals are dispersive. Also, it becomes cumbersome to set up all the required simultaneous equations. Therefore this problem is left to Chapter 5 where the matrix methods are introduced to handle just such situations.

Analysis of Non-Collinear Beams

The full analysis of the arbitrary multi-member joint is rather cumbersome and only the special case of the non-collinear beam is treated here. More detailed results on the general case can be found in References [58,64]. The angled joint is important because it introduces the idea of mode conversion. That is, incident flexural waves can generate longitudinal waves and *vice versa*. It is also pursued here as a precursor to the exact analysis of curved beams which follows in the next section.

The equations of motion of the three member joint in Figure 4.7 are

$$
\begin{aligned}
-F_1 + F_2 \cos\theta - V_2 \sin\theta &= m\ddot{u}_J \\
-V_1 + F_2 \sin\theta + V_2 \cos\theta &= m\ddot{v}_J \\
-M_1 + M_2 + \tfrac{1}{2}L(V_1 + V_2) &= I\ddot{\phi}_J
\end{aligned}
\tag{4.22}
$$

where L is a size estimate of the joint. Displacement continuity for small joint rotations are

$$
\begin{aligned}
u_1 &= u_J \\
u_2 &= u_J \cos\theta + v_J \sin\theta
\end{aligned}
$$

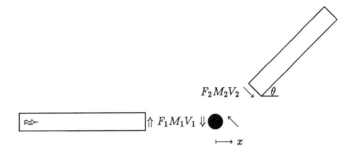

Figure 4.7: Joint boundary conditions.

$$v_1 = v_J - \tfrac{1}{2}L\phi_J$$
$$v_2 = -u_J \sin\theta + v_J \cos\theta + \tfrac{1}{2}L\phi_J$$

Slope continuity conditions are

$$\frac{\partial v_1}{\partial x} = \frac{\partial v_2}{\partial x} = \phi_J \tag{4.23}$$

The axial force, shear force and bending moment, respectively, are related to the displacements by

$$F = EA\frac{\partial u}{\partial x}, \qquad V = -EI\frac{\partial^3 v}{\partial x^3}, \qquad M = EI\frac{\partial^2 v}{\partial x^2}$$

Finally, if the flexural displacements are expanded as equation (4.5) and the longitudinal displacements as

$$u_1(x,t) = \sum_n [\mathbf{P}_n e^{-k_{Ln1}x} + \mathbf{Q}_n e^{+k_{Ln1}x}]e^{i\omega_n t}$$
$$u_2(x,t) = \sum_n [\bar{\mathbf{P}}_n e^{-ik_{Ln2}x}]e^{i\omega_n t} \tag{4.24}$$

where k_{Ln} is the longitudinal wavenumber, then a system of six equations can be set up so as to determine the six wave coefficients. Such a system of equations is frequency dependent. This comes about not only from the mass but also from the presence of mixed wavenumber terms. For example, both V and F occur in the same equation, and since their wavenumbers have different dependencies on frequencies, then the frequency does not cancel through.

Consider, as a special case, when both beam segments are similar but just ori-

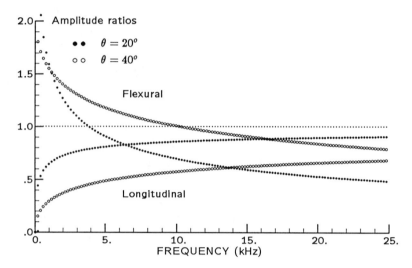

Figure 4.8: The effect of frequency on the transmitted amplitude ratios.

ented differently. Then the system of equations become

$$
\begin{bmatrix}
1 & 1 & -\cos\theta & -\cos\theta & 0 & -\sin\theta \\
0 & 0 & \sin\theta & -\sin\theta & 1 & -\cos\theta \\
i & 1 & i & 1 & 0 & \\
0 & 0 & ik^3K\sin\theta & -k^3K\sin\theta & -ik_L & -ik_L\cos\theta \\
-k^3K & k^3K & -ik^3K\cos\theta & k^3K\cos\theta & 0 & -ik_L\sin\theta \\
1 & -1 & -1 & 1 & 0 & 0
\end{bmatrix}
\begin{Bmatrix}
C \\ D \\ \hat{A} \\ \hat{B} \\ Q \\ \hat{P}
\end{Bmatrix}
$$

$$
= \begin{Bmatrix} -1 \\ 0 \\ i \\ 0 \\ -ik^3K \\ -1 \end{Bmatrix} A +
\begin{Bmatrix} 0 \\ -1 \\ 0 \\ ik_L \\ 0 \\ 0 \end{Bmatrix} P \qquad (4.25)
$$

where $K \equiv EI/EA$. The presence of the axial terms means there are longitudinal waves generated in each beam section.

Figure 4.8 shows the variation of the transmitted flexural and longitudinal amplitude ratios with frequency when a longitudinal wave is incident. Similar results are obtained for an incident flexural wave. These plots show that the joint has an obvious frequency effect. Irrespective of the angle, the higher-frequency components tend to be unaffected by the angle change. That is, the amplitude ratio tends to unity. Conversely, for the generated flexural wave, the amplitude tends to zero at high frequencies. Interestingly, even for small angles, the joint acts as a frequency filter for the low frequencies and as a result a large flexural component is always generated. Actually, the particular limits attained depend on the cross-sectional properties. The reflected amplitude ratios, shown in Figure 4.9, illustrate how a flexural amplitude is also reflected.

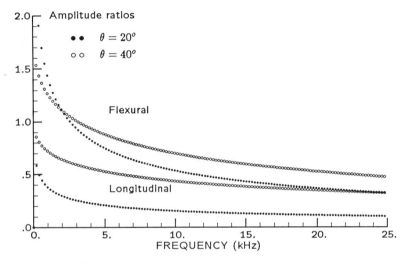

Figure 4.9: The effect of frequency on the reflected amplitude ratios.

If a curved rod is considered as a sequence of angled joints, then similar results are expected there also. It must be borne in mind in both of these examples that beyond 100kHz or so the Bernoulli-Euler beam theory is no longer valid and the improved Timoshenko theory should be used.

4.5   Curved Beams

A companion problem to waves in a jointed beam is that of the curved beam or rod. There is considerable intrinsic interest in these because of such structural applications as arches and helical springs. More detailed analysis than what will follow can be found in References [31,95,165,182,232]. There is also the special case of negligible bending stiffness which corresponds to waves in cables and powerlines. Some interesting analyses of this can be found in References [149].

The interest here, however, is to introduce a situation where the spectrum relation may be difficult to solve in closed form and where numerical methods must be resorted to.

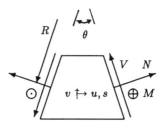

Figure 4.10: Free body diagram for curved beam element.

Equations of Motion

Let the rod be narrow and have a fairly large radius of curvature, then with respect to Figure 4.10, the resultants are the normal force N, the shear force V, and the bending moment M. The equations of motion are

$$\frac{\partial N}{\partial s} - \frac{V}{R} = \rho A \frac{\partial^2 u}{\partial t^2}, \qquad \frac{\partial V}{\partial s} + \frac{N}{R} = \rho A \frac{\partial^2 v}{\partial t^2}, \qquad \frac{\partial M}{\partial s} + V = 0 \qquad (4.26)$$

where s is the arc length related to the radius R by $s = R\theta$. These relations are valid for arbitrary plane curved beams; only if R is constant are they restricted to circular ones. Obviously if R becomes very large the straight beam equations (4.1 & 4.2) are recovered.

The resultants N and M are related to the deformation by

$$N = EA \left[\frac{\partial u}{\partial s} - \frac{v}{R} \right], \qquad M = EI \frac{\partial}{\partial s} \left[\frac{\partial v}{\partial s} - \frac{u}{R} \right] \qquad (4.27)$$

Combining these with the equations of motion gives the coupled set of equations (for $R = $ constant)

$$EA \frac{\partial^2 u}{\partial s^2} \left(1 - \frac{\beta}{R^2} \right) + \frac{EA}{R} \left\{ -\frac{\partial v}{\partial s} + \beta \frac{\partial^3 v}{\partial s^3} \right\} = \rho A \frac{\partial^2 u}{\partial t^2}$$

$$EI \frac{\partial^4 v}{\partial s^4} + \frac{EI}{R} \left\{ \frac{\alpha}{R} v - \frac{\partial^3 u}{\partial s^3} - \alpha \frac{\partial u}{\partial s} \right\} = -\rho A \frac{\partial^2 v}{\partial t^2} \qquad (4.28)$$

where $\alpha = EA/EI$ and β is its reciprocal. Again, the equations for the rod and beam can be recovered by putting $R = \infty$. The terms in the braces are the coupling between the flexural and axial motions. Obviously, if $R = \infty$, there is no coupling.

Spectrum Relation

Since the coefficients are constant, then the spectrum relation is obtained by first assuming solutions of the form

$$u(s,t) = \sum \hat{u}_o e^{-i(ks - \omega t)}, \qquad v(s,t) = \sum \hat{v}_o e^{-i(ks - \omega t)} \qquad (4.29)$$

On substitution into equation (4.28) this gives

$$
\begin{bmatrix}
EAk^2\left(1-\beta/R^2\right)-\rho A\omega^2 & -ikEA(1+\beta k^2)/R \\
ikEI(\alpha-k^2)/R & EI\left(k^4+\alpha/R^2\right)-\rho A\omega^2
\end{bmatrix}
\begin{Bmatrix} \hat{u}_o \\ \hat{v}_o \end{Bmatrix} = 0
\qquad (4.30)
$$

The characteristic equation to determine the wavenumber k is then formed by setting the determinant of this system to zero. The resulting equation is

$$
k^6 - k^4\left\{k_L^2\right\} - k^2\left\{\frac{1}{R^4}+k_F^4\left(1-\frac{\beta}{R^2}\right)\right\} + k_L^2\left\{-\frac{\alpha}{R^2}+k_F^4\right\} = 0
\qquad (4.31)
$$

where $k_L \equiv \omega[\rho A/EA]^{1/2}$ and $k_F \equiv \sqrt{\omega}[\rho A/EI]^{1/4}$ are the wavenumbers for the pure longitudinal and flexural modes, respectively. This has six solutions in all, but since it is cubic in k^2, then there are three basic modes: one associated with the longitudinal behavior and two associated with the flexural. This can be seen by noting that for very large R the above can be factored into

$$
(k^2 - k_F^2)(k^2 + k_F^2)(k^2 - k_L^2) = 0
$$

In general, of course, the modes are coupled and it is not proper to speak of a longitudinal mode or a flexural mode. Also note that when the frequency is zero, it can be factored as

$$
k^2(k^2 - \frac{1}{R^2})(k^2 + \frac{1}{R^2}) = 0
$$

and only on root goes through zero.

It can be pointed out at this stage that, although the spectrum relation may appear in a complicated transcendental form, this does not diminish the spectral approach. This case should be tackled in a two-step manner. First the spectrum relation is solved numerically for the particular material, geometry, and so on of interest. It is then curve fitted or tabulated to be utilized in the propagating programs. Obtaining the spectrum relation can be a big undertaking in itself and that is why every opportunity is taken in this book to extract as much information as possible from it before actually solving for the responses. Indeed, with a little experience, only the spectrum relation and the transfer function are all that will be required.

Solving for $k(\omega)$

The first step is to non-dimensionalize or parameterize the characteristic equation as much as possible. In the present case, introduce the non-dimensional quantities

$$
\bar{k} \equiv kh, \qquad \xi \equiv \frac{h}{R}, \qquad \bar{\omega} \equiv \frac{\omega h}{c_o}
$$

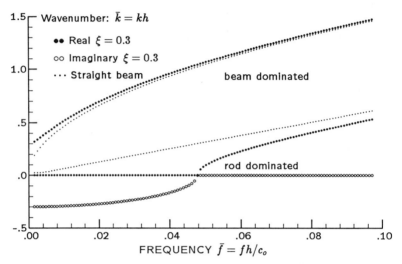

Figure 4.11: Spectrum relations for curved beams.

where h is the beam thickness. Noting that $\alpha\beta = 1$ allows

$$\bar{k}^6 - \bar{k}^4\left\{\bar{\omega}^2\right\} - \bar{k}^2\left\{\xi^4 + \bar{\omega}^2(12 - \xi^2)\right\} - 12\bar{\omega}^2\left\{\xi^2 - \bar{\omega}^2\right\} = 0 \qquad (4.32)$$

Thus the normalized spectrum relation $\bar{k}(\bar{\omega})$ is dependent on the single parameter ξ since $\bar{\alpha} = h^2 EA/EI = 12$ for a rectangle. This equation can be written as a polynomial in the form

$$z^3 + a_2 z^2 + a_1 z + a_0 = 0$$

where $z \equiv k^2$ and the a_n have obvious meanings. This particular equation can be solved in closed form but it is now desired instead to introduce a numerical scheme for achieving this.

The idea of the scheme is to consider the complex function above as two separate real functions in (x, y) space. That is, the roots occur where

$$f(z) = g(x, y) + ih(x, y) = 0$$

For $f(z)$ to be zero, it is necessary that both g, h be zero simultaneously. That is, it is necessary to find those (x, y) pairs where these functions are zero. So introduce the auxiliary real function that embodies this. A simple example is

$$w(x, y) = |g(x, y)| + |h(x, y)|$$

which is just the sum of the absolute values. The problem is thus reduced to one of finding the minimum value of the real function $w(x, y)$ in the real (x, y) space.

A straightforward scheme for finding the minimum of a function is to start at some point (x_0, y_0), say, and by estimating the slopes there (by function evaluations at neighboring points (x_1, y_1), (x_2, y_2)) a direction for marching toward a minimum can be chosen. This is repeated many times with decreasing step size until a function value as small as desired is achieved. A version of this, coded in FORTRAN, is given in the program ROOTCOMP in Appendix D. This program has a few extra features in order to check convergence and to choose the step size, but the algorithm is essentially as laid out above. As can be seen, the program need not be very large.

There are no hard and fast rules about obtaining good initial guesses. However, once a root is obtained, then this can be used as the initial guess for the next one of the same spectrum. There is generally no problem until two spectra overlap or come very close to each other. To clarify ambiguities in these cases it is useful to approach the coupled case incrementally from the uncoupled situation. It is reiterated that the scheme outlined above is not used "on the fly" as part of the propagation problem, but, instead, is a separate analysis. Thus there is room for trial and error, and it is not necessary that the algorithm be foolproof for every situation. Once a single spectrum has been determined, then the others can be obtained in a similar manner (by starting with a smart guess) or by using a factorization scheme. This depends on the particular problem.

Figure 4.11 shows two modes of the spectrum relation for the curved beam. The coupling, generally, will not cause drastic changes in the spectrums. This is clear from the figure. As regards the specifics of these plots, notice how the spectra are relatively unaffected at the higher frequencies. This is a conclusion similar to that for the angled joint. The particular deviation from the uncoupled theory corresponds to the effect due to an axial compression. The consequences of this for beams are treated later in this chapter. For the rod it can be seen that the lower frequencies do not propagate. There is a cut-off frequency at

$$\omega_o = \frac{c_o}{R}$$

Another possibility is to use matrix methods to avoid setting up an analytically coupled situation. For example, the curved beam can be treated as a collection of small straight segments. Each segment then has a simple, well-defined spectrum relation — the complexity associated with the curved beam will come from the interaction of all the small segments. This approach is shown in the next chapter.

4.6   Remote Sensing

There are many problems in engineering mechanics that require the monitoring of a structure for the occurrence of some sudden event. The remote sensing of a space platform, for example, would require knowing if the platform was struck (by a meteor say), where it was struck, and how large the force was. Another example is the use of acoustic emission to locate the position of a flaw by triangulating on the received pulses (Reference [208]). Ultrasonic detection techniques use transducers to both send and receive waves. By examining the distortion in the returning signal it is possible to locate the presence of flaws (Reference [106]). Similar issues arise in the use of ultrasonic imaging (Reference [146]).

The purpose of this section is to show how a spectral analysis of dispersive signals monitored at two locations can be used to locate the source and determine the original shape of the signal. The approach focuses on the utilization of the phase behavior. The specific example of flexural waves in beams originating from an impact is considered but the procedures can be applied more generally. It must be emphasized that this is an experimental problem — inverse problems use experimental data as input to determine the excitation signal.

Background

The bulk of the work in this area of remote sensing has concentrated on non-dispersive waves (or at least the non-dispersive part of waves). Since these show little or no change in shape as they propagate, they have a simple phase relationship given by

$$\text{phase} = (x - c_o t) \tag{4.33}$$

where c_o is the constant wave speed. Consequently, it is easy to identify and keep track of the pulses in space and time. For example, in seismology (Reference [81]) the origin of an earthquake can be determined by noting the relative arrival times of the P, S and Rayleigh waves, since each of these travel at a constant (but different) speed.

A more complicated range of problems arise when the waves are dispersive. Here the ability to identify the wave as it propagates is difficult because its shape changes and the phase speed is frequency dependent. The research to date dealing with remote sensing in dispersive media is quite limited and undeveloped; the following references give an idea of some of the approaches taken. Reference [107] showed that the use of transient toneburst sequences (these are essentially single-frequency excitations) permits the measurement of group velocities and attenuation. However, the cross-correlation schemes do not seem to be successful when an arbitrary excitation is used. Reference [29] discusses how dispersed signals can be recompressed to give sharp arrival times. The scheme introduced is based on the conversion of the frequency transform of the signal to a wavelength transform as originally introduced for imaging faults in coal seams (Reference [27]). This method cannot tell absolute positions, but if there are multiple reflections, then it can tell the distance between boundaries. A different approach entirely is taken in References [114,230]. First an estimate of the position is obtained from the product of the transient duration by an average phase speed. Both of these are estimates obtained from the signal at a remote location. The position is then iterated on until the reconstructed force has no significant negative portions. This approach is adequate for impact-type forcing histories, but is unsuitable for detecting the origin of a general disturbance. For example, it would not work for the thermal blast of Chapter 3. In all the above cases, a flexural wave in a beam was used as the source of the dispersive signals.

Spectral Analysis of Recorded Signals

Consider the experimental arrangement of a long aluminum beam instrumented with strain gages and impacted with a steel ball. The exact positions of the gage pairs (relative to the impact site) are given in Table 4.2 and details of the experimentation can be found in Reference [66]. Note that the only function of the impact itself is to act as an unknown spectrum source.

Gage combination	Calculated Position from each gage				Standard Deviation	Differences	% of Window
B C'	1.80	4.49	-	-	± 0.47	0.05	0.8%
B D'	1.49	-	8.76	-	± 0.55	0.26	2.5%
B E'	1.55	-	-	12.70	± 0.63	0.20	1.4%
C D'	-	4.17	8.83	-	± 0.67	0.33	2.5%
C E'	-	4.18	-	12.82	± 0.91	0.32	1.9%
D E'	-	-	8.56	12.44	± 0.47	0.06	0.3%
actual =	1.75	4.50	8.50	12.50			

Table 4.2: Summary of impact position predictions. All distances in mm.

During each test, signals from strain gage pairs straddling the impact site are recorded and Figure 2.10 shows some of the responses from gages (B, C') and (D, E'). These sets are quite typical of the other gage arrangements. As demonstrated in the previous sections, in order to get good spectral estimates from finite beams it is necessary to both append the measured history and pad with zeros. The resulting appended traces are the ones used in the following analysis.

The strain time histories are converted to frequency data by using a 512 point FFT transform with a sampling rate of 5 μs. This corresponds to a time window of 2560 μs and a Nyquist frequency of 100 kHz. At each frequency ω_n, the strain transform (for a given gage) can be written as

$$\hat{e}(x, \omega_n) = a_n + ib_n = C_n e^{i\phi_n}, \qquad\qquad i = \sqrt{-1} \qquad (4.34)$$

where a_n and b_n are the real and imaginary parts, respectively, of the transform. Alternatively, the transform can be expressed in terms of its amplitude C_n and phase ϕ_n. The amplitude spectra of the strain traces are unremarkable, and they overlay quite consistently except, as expected, in the vicinity of very low or very high frequency. It is noteworthy that beyond the 20-30 kHz range the amplitudes are negligible.

Figure 4.12 shows the phase behaviors of the trace at $x = 4.5$in. The saw-tooth pattern is due to the ambiguity when using the arctangent function to obtain the phase by

$$\phi = \arctan[b/a] \pm N\pi \qquad (4.35)$$

To utilize the phase information it is necessary to remove this ambiguity by "unwrapping" the phase. While it is possible to manually adjust by $N\pi$ or to write computer software to do this (Reference [219] gives software used for the analogous problem in ceptrum analysis), both are rather cumbersome. Instead the following simpler, but adequate, scheme is used. Much of the saw-tooth pattern can be unwrapped by simply evaluating the phase as

$$\phi_n = \arctan\left[b_n^*/a_n^*\right] + k_n x^* \qquad (4.36)$$

where

$$\left(a_n^* + i b_n^*\right) = \left(a_n + i b_n\right) e^{i k_n x^*}$$

That is, the signal is imagined to be propagated back a distance x^* to where the phases are relatively small. Performing the arctangent will then give fewer discontinuities. The rest of the equation just replaces the phase that was removed. The improvement is shown in Figure 4.12 for a series of guesses of x^*. Any reasonable value of x^* will help in removing the discontinuities, and as part of an iteration scheme the guess can be changed to give the maximum improvement. Note that this approach works because information about the structure in the form of $k(\omega)$ is known. This scheme could not be used directly on a signal in an arbitrary unknown system. If the time origin of the pulse is also unknown, then there are additional initial phases due to the $e^{i\omega t_o}$ term. Therefore, the above scheme can never completely unwrap the phase, but coupled with the following procedures it works effectively.

Relation Between Phase and Position for Flexural Waves

The response for a beam can be put in a form that emphasizes the phase relationships as

$$C_n e^{i\phi_n} = C_{on} e^{i\phi_o}[D_n e^{i\theta_n}]e^{-i k_n x}$$

The term inside the square bracket is associated with the "ringing" of the beam, that is, the spatially damped vibration generated when flexural waves interact with discontinuities and is given by

$$[D_n e^{i\theta_n}] \equiv [1 + i e^{i k_n x} e^{-k_n x}]$$

Its presence makes the phase relation nonlinear in x, but this function is significant only for small x (i.e., close to the signal initiation site) and for low frequencies. Consequently, the relatively minor behavior of the ringing term allows it to be approximated with any reasonable guess for x. Further, if an iteration scheme is

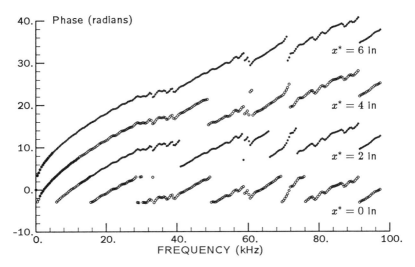

Figure 4.12: Improvements achieved in unwrapping the phase.

also used, then the guess can be improved and so the effect of this function can be included exactly.

The relation between phase and position can now be written as (recalling that θ_n is assumed known at each iteration)

$$\phi_n = \phi_{on} + \theta_n - k_n x \qquad (4.37)$$

This equation forms the basis of the method to obtain x. Ideally, it should be possible to obtain the position from the phase by rearranging this equation to give

$$x = \frac{1}{k_n} \left[\phi_{on} - \phi_n + \theta_n \right]$$

The saw-tooth pattern of Figure 4.12 prevents this, however, because it is not always possible to unwrap all the jumps in phase. It is easier instead to determine x from a derivative of equation (4.37), that is,

$$x = \frac{d}{dk_n} \left[\phi_{on} - \phi_n + \theta_n \right] \qquad (4.38)$$

For ideal data this would give a constant value of x except at the points of discontinuity in phase, and if these points are kept to a minimum, then there is little difficulty in interpreting the results. Notice that the derivative is independent of the absolute value of the phase shift $N\pi$. (In fact, the shift is only a factor if the derivative is taken across a jump.)

When experimental data are used in conjunction with the finite difference method
to obtain the derivative, a large variation in x can be found. This is shown in Figure
4.13 where a central difference is used for the derivative. Many schemes can be used
to improve the derivative, such as using an n-point average central difference where
n is chosen depending on the resolution of the data. Alternatively, since only a single value of x is required and since a separate value is obtained at each frequency,
then sufficient data are available to do a statistical estimate of x. For example,
simply computing the average of all the estimates can give good results. Note that
the scheme is made more robust by discarding all values outside a predetermined
range (i.e., one standard deviation say.) Keep in mind that only a single value of
position is required and so there are many possibilities for performing statistics on
the multiple values produced.

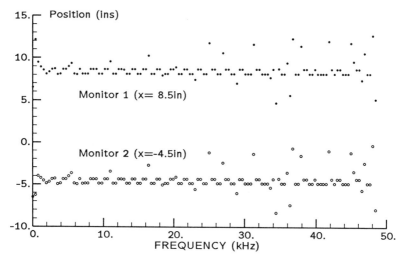

Figure 4.13: Typical estimates of position as a function of frequency.

Locating the Space Origin of a Pulse

The method requires that there be two sensors, one on each side of the origin of the
pulse. The phase/position relation for each sensor is

$$x_1 = \frac{d}{dk_n}[\phi_{on} - \phi_{1n} + \theta_{1n}]$$

$$x_2 = \frac{d}{dk_n}[\phi_{on} - \phi_{2n} + \theta_{2n}] \tag{4.39}$$

It is the unknown initial phase ϕ_n that prevents the use of only a single sensor. But
since both of the above relations have the same initial phase, then they constitute

a set of two equations with three unknowns (x_1, x_2, ϕ_{on}). The additional relation is obtained because the absolute spacing between the sensors is also required to be known, that is,

$$x_1 + x_2 = L = \text{ known}$$

then the initial phase can be removed from the above equations to give

$$x_1 = \tfrac{1}{2}L - \tfrac{1}{2}\frac{d}{dk_n}\left[(\phi_{1n} - \phi_{2n}) - (\theta_{1n} - \theta_{2n})\right]$$

$$x_2 = \tfrac{1}{2}L + \tfrac{1}{2}\frac{d}{dk_n}\left[(\phi_{1n} - \phi_{2n}) - (\theta_{1n} - \theta_{2n})\right] \tag{4.40}$$

Recalling that good estimates can be obtained for θ by assuming reasonable guesses for x, then the above can be used for estimating the source of a pulse from measured phase differences of two sensors. Incorporating these equations as part of an iteration scheme improves the accuracy and consistency.

The responses from various combinations of gages were recorded for nominally the same impact and a set of predicted positions obtained. The means and average standard deviations of these are given in Table 4.2. The means are quite close to the actual. In all cases, the iteration was begun by assuming the impact to be at the center of the window and convergence was achieved within 2 to 3 iterations. The standard deviation is a measure of the variability of the derivative and can be monitored so as to decide when to use more smoothing or averaging.

It is worth pointing out that the sensors must straddle the source, otherwise they would only contain information about the phase change incurred by the signal in propagating from one sensor to the other. Also, it might appear that one sensor should be adequate since each frequency component travels at a different speed. Unfortunately, this does not work because each frequency component also has a different unknown initial phase.

Determining the Time Origin of a Pulse

The triggering of the recording of the sensor histories does not necessarily coincide with the beginning of the initiation event. That is, the time origin of the pulse is also unknown. The simplest method to determine the time origin is to reconstruct the pulse in the time window. Using equation (4.11) and rearranging it for force identification from strain, it can be shown that the frequency components of the force can be obtained from the frequency components of the strain by

$$\hat{F}_n = 8EI\frac{h}{k_n}\hat{e}(x, \omega_n)/\left[ie^{-ik_n z} - e^{-k_n z}\right] \tag{4.41}$$

When this is reconstructed in the time domain by use of the inverse transform, then the complete history of the pulse is obtained. From this, the time origin will be

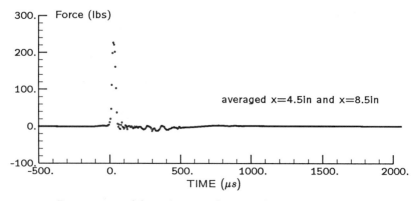

Figure 4.14: Reconstructed force history showing the time origin of the source.

apparent. Figure 4.14 shows a typical reconstructed force history. It is seen that the pulse is reconstructed in the correct time location. On a time scale on the order of the recorded strain (i.e., about 600 μs) all pulses appear to be located at the same origin.

4.7 General Bernoulli-Euler Beam

Once the basic beam is understood, it is a straightforward matter to enrich the available solutions by considering such features as viscoelasticity and variable thickness. Actually, viscoelasticity will not be considered here because, just as for the rod, the only difference is that it is slightly more complicated by the fact that the modulus E is now a function of frequency. But since the solution is dispersive to begin with, then there is no fundamental change in the appearance of the waves. Reference [168] contains analysis, both theoretical and experimental, of the vibration of a viscoelastic beam. The effect of a viscoelastic layer in a sandwich beam is considered in References [150,203]. Of interest here are the consequences of pre-stress and variable dimensions on the beam behavior.

Beam on a Foundation

Let the beam be on a foundation of elastic stiffness K, of viscosity η and let it have a pre-tension T. The differential equation of motion for the Winkler kind of foundation is

$$EI\frac{\partial^4 v}{\partial x^4} + \rho A\frac{\partial^2 v}{\partial t^2} + Kv + \eta\frac{\partial v}{\partial t} - T\frac{\partial^2 v}{\partial x^2} = P \qquad (4.42)$$

This equation has some interesting variations on the usual beam solutions. Since the coefficients are constant, then the spectrum relation can be obtained simply by

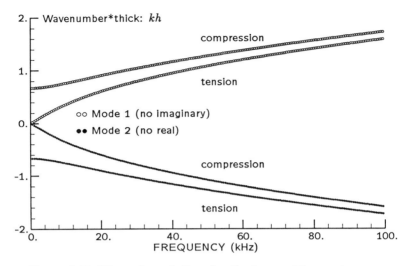

Figure 4.15: Dispersion relations for the beam with pre-stress.

taking solutions of the form $v(x,t) = v_0 e^{-i(kx-\omega t)}$ to get

$$EIk^4 - \rho A\omega^2 + K + i\omega\eta + Tk^2 = 0$$

giving the wavenumbers as

$$\sqrt{2EI}\, k = \pm \left[-T \pm \sqrt{T^2 + 4EI(\rho A\omega^2 - K - i\omega\eta)} \right]^{\frac{1}{2}} \qquad (4.43)$$

There are a total of four solutions, as usual, and this is unchanged by the addition of the new features. Rather than discuss this in general, it is more instructive to consider some of the special cases separately.

Consider, for example, the case of a free standing beam with initial stress T, then $K = 0$ and $\eta = 0$ giving

$$\sqrt{2EI}\, k = \pm \left[-T \pm \sqrt{T^2 + 4EI\rho A\omega^2} \right]^{\frac{1}{2}} \qquad (4.44)$$

The sign of the pre-stress can make a significant difference to the character of the solution as shown in Figure 4.15. The interpretation of this diagram is that for the propagating mode the wavelengths in the compressed beam are decreased (the wavenumber is increased), while those in the stretched beam are extended compared to the unstressed beam. The effect of this on the propagating speeds is shown in Figure 4.16. Notice that the group speed is increased. This situation is similar to

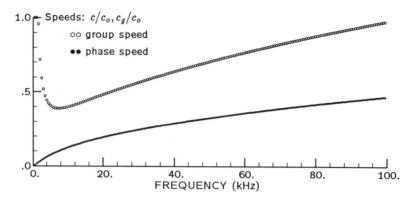

Figure 4.16: Dispersion relations for the beam with compressive pre-stress.

the case of the curved beam because there the coupling is an axial compression. Further, only wavelengths less than $2\pi EI/T$ can exist in the compressed beam.

A beam on an elastic foundation is modeled by taking $\eta = 0$ and $T = 0$. The spectrum relations are then

$$k_1 = \pm \left[\frac{\rho A}{EI}\omega^2 - \frac{K}{EI}\right]^{1/4} \, , \qquad k_2 = \pm i \left[\frac{\rho A}{EI}\omega^2 - \frac{K}{EI}\right]^{1/4} \qquad (4.45)$$

As usual the wavenumber will have both real and imaginary roots but the form suggests the possibility of a cut-off frequency. Look for this cutoff by finding the frequency where k is zero, that is

$$\frac{\rho A}{EI}\omega^2 - \frac{K}{EI} = 0 \qquad \text{or} \qquad \omega_o = \sqrt{\frac{K}{\rho A}} \qquad (4.46)$$

The spectrum relations for the four modes can be written as

$$k_1 = \pm \left[\frac{\rho A}{EI}(\omega^2 - \omega_o^2)\right]^{1/4} \, , \qquad k_2 = \pm i \left[\frac{\rho A}{EI}(\omega^2 - \omega_o^2)\right]^{1/4} \qquad (4.47)$$

Then for frequencies greater than ω_o there are two purely real and two purely imaginary roots as is usual for beams. However, for frequencies less than ω_o

$$k = \pm \frac{1}{\sqrt{2}}(1 \pm i)\left[\frac{\rho A}{EI}(\omega_o^2 - \omega^2)\right]^{1/4} \qquad (4.48)$$

which shows that all roots are complex. Thus all components associated with both modes attenuate to some extent. This situation is shown in Figure 4.17. Of the available solutions, it is necessary to choose those that decrease in the direction of propagation. Consequently, there is a standing wave set up at frequencies less than ω_o. The concept of speed has no meaning in this case. The Bernoulli-Euler beam solution is recovered simply by putting $\omega_o = 0$. Note that $\omega = \omega_o$ is a branch point and it is difficult to determine the continuation of each mode. This ambiguity is removed by adding a little damping as discussed previously.

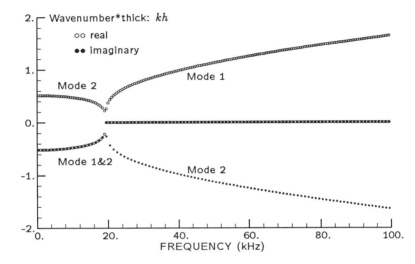

Figure 4.17: Dispersion relations for the beam on an elastic foundation.

Reference [202] considers the interesting problem of two parallel beams connected by elastic springs while the damping effect of beams vibrating in a fluid is considered in Reference [111]. Both of these give additional insights into the results of this section.

Beam of Variable Thickness

The following example is restricted to a linear thickness variation since this captures the essence of the effects. A more general analysis can be found in Reference [89].

Suppose the taper in the beam is only slight, then it is reasonable to assume that the equations of motion as derived are still valid and that the only change is to consider the area as a function of position. That is, in spectral form

$$\frac{d^2}{dx^2}\left[EI(x)\frac{d^2\hat{v}}{dx^2}\right] - \omega^2 \rho A(x)\hat{v} = 0 \tag{4.49}$$

By way of example let the thickness of the beam be given by the form

$$h = h_o \left(\frac{a+x}{a}\right)^m \tag{4.50}$$

so that the area and moment of inertia become

$$A = A_o \left(\frac{a+x}{a}\right)^m , \qquad I = I_o \left(\frac{a+x}{a}\right)^{3m}$$

since the width is assumed constant. The definition of the geometry is similar to that given in Figure 3.11. When this is substituted into the differential equation, then get

$$(a+x)^{2m}\frac{d^4\hat{v}}{dx^4} + 6m(a+x)^{2m-1}\frac{d^3\hat{v}}{dx^3} + 3m(3m-1)(a+x)^{2m-2}\frac{d^2\hat{v}}{dx^2} - a^{2m}\beta^4\hat{v} = 0$$

where $\beta^4 = \omega^2(\rho A_o/EI_o)$. Consider the special case when $m = 1$, then this can be factored into product operators as

$$\left[(a+x)\frac{d^2}{dx^2} + 2\frac{d}{dx} + a\beta^2\right]\left[(a+x)\frac{d^2}{dx^2} + 2\frac{d}{dx} - a\beta^2\right]\hat{v} = 0$$

Consequently, independent solutions are obtained by solving, separately,

$$(a+x)\frac{d^2\hat{v}}{dx^2} + 2\frac{d\hat{v}}{dx} - a\beta^2\hat{v} = 0, \qquad (a+x)\frac{d^2\hat{v}}{dx^2} + 2\frac{d\hat{v}}{dx} + a\beta^2\hat{v} = 0$$

This is recognized (from Appendix B) as a set of Bessels equations having the solutions

$$\hat{v}(x) = \frac{1}{\sqrt{z}}\{AJ_2(z) + BY_2(z) + CI_2(z) + DK_2(z)\} \tag{4.51}$$

where the argument is $z = 2\beta\sqrt{a}\sqrt{a+x}$. In terms of the forward moving wave it is seen that it decreases in intensity. In this result there are two sources of dispersion, one due to the spectrum relation and the other due to the Bessel functions.

The more complicated the area variation, the less likely it is that a generic solution can be found. A different tack, then, is to take a hint from the finite element method approaches and replace the continuous variation of section properties by a stepwise constant (or other simple) variation. This replaces the single complicated differential equation by multiple simpler ones. The matrix methods developed in the following chapter make this a feasible alternative.

4.8   Timoshenko Beam Theory

The elementary beam theory predicts unrealistic speeds at high frequencies. Looking at the assumptions in deriving this theory, neglecting the rotational inertia is an obvious failing. While the magnitude of ρI may be small, at very high frequencies the contribution of $\ddot{\phi}$ could be significant. A subtler assumption used is that the shear deformation is zero (even though there is a shear force). Both of these assumptions will be relaxed in the present higher-order theory.

Equations of Motion

The equations of motion in terms of the resultants are the same as before except that the rotational inertia is not neglected. That is

$$\frac{\partial V}{\partial x} = \rho A \frac{\partial^2 v}{\partial t^2} - P$$
$$V + \frac{\partial M}{\partial x} = \rho I \frac{\partial^2 \phi}{\partial t^2} \tag{4.52}$$

The axial and transverse displacements are taken to be

$$u(x, y, t) = -y\phi(x, t), \qquad v(x, y, t) = v(x, t) \tag{4.53}$$

giving the strains as

$$e_x = \frac{\partial u}{\partial x} = -y \frac{\partial \phi}{\partial x}$$
$$e_y = \frac{\partial v}{\partial y} = 0$$
$$2e_{xy} = \frac{\partial u}{\partial y} + \frac{\partial v}{\partial x} = -\phi + \frac{\partial v}{\partial x} \tag{4.54}$$

Thus the deformation is described by the two independent functions $v(x, t)$ and $\phi(x, t)$. Note that only the axial strain is a function of y. From the third of these equations, it is observed that if the slope and deflection are related by $\phi = \partial v / \partial x$, then there is zero shear strain. This is the assumption in the elementary theory and will not be imposed here. Also note that this behavior is unrelated to the shear force occurring in the equation of motion. That is, both shears arise in the equations for different reasons.

The stresses and consequent resultants are obtained by first assuming the material is linear elastic giving the stresses as

$$\sigma_x = Ee_x = -Ey \frac{\partial \phi}{\partial x}$$
$$\sigma_y = 0$$
$$\tau_{xy} = G2e_{xy} = G \left[\frac{\partial v}{\partial x} - \phi \right] \tag{4.55}$$

Noting that only σ_x is a function of y, then the resultants are obtained by integration on the cross-section to give

$$V = \int \tau_{xy} \, bdy = GAK \left[\frac{\partial v}{\partial x} - \phi \right]$$
$$M = -\int \sigma_x y \, bdy = EI \frac{\partial \phi}{\partial x} \tag{4.56}$$

where b is the beam width. The coefficient K depends on section properties and is introduced as a correction factor since the distributed shear strain is not constant on the section.

The so-called Timoshenko equations are obtained by taking v and ϕ as the dependent variables giving

$$GAK\left[\frac{\partial^2 v}{\partial x^2} - \frac{\partial \phi}{\partial x}\right] = \rho A\frac{\partial^2 v}{\partial t^2} - P$$

$$EI\frac{\partial^2 \phi}{\partial x^2} + GAK\left[\frac{\partial v}{\partial x} - \phi\right] = \rho I\frac{\partial^2 \phi}{\partial t^2} \qquad (4.57)$$

These are simultaneous second-order differential equations in v and ϕ. It is possible to reduce this to a single differential equation in the deflection, but it is more usual to leave it in its present form. The group constant GAK is essentially the shear stiffness. While elastic materials have a definite relation between G and E, here they are allowed to be distinct. Indeed, in order to recover the elementary theory, it is necessary to set $GAK = \infty$. This, of course, is saying there is no shear deformation even though there are deformations present. The special cases of interest are when rotational effects are neglected (i.e., $\rho I = 0$) and when shear deformations are neglected (i.e., $G = \infty$). These will be looked at in greater detail later.

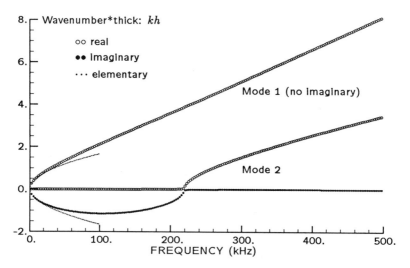

Figure 4.18: Spectrum relations for the Timoshenko beam theory.

Spectrum Relation

Since there are two dependent variables v, ϕ and the coefficients are constant then assume

$$v = v_o e^{-i(kx-\omega t)}, \qquad \phi = \phi_o e^{-i(kx-\omega t)}$$

Substituting into the Timoshenko equations gives

$$\begin{bmatrix} GAKk^2 - \rho A\omega^2 & -ikGAK \\ ikGAK & EIk^2 + GAK - \rho I\omega^2 \end{bmatrix} \begin{Bmatrix} v_0 \\ \phi_0 \end{Bmatrix} = 0 \qquad (4.58)$$

Define, for convenience, the new constants

$$c_o q \equiv \sqrt{\frac{EI}{\rho A}}, \qquad c_s \equiv \sqrt{\frac{GAK}{\rho A}}, \qquad Q \equiv \sqrt{\frac{\rho I}{\rho A}}$$

In terms of geometry, I and A have a simple relationship and so it might be thought that Q can be simplified. But in the present analysis ρI is associated only with the rotational inertia, and by keeping it distinct from ρA it will be possible to set it to zero. The above becomes in terms of these new constants

$$\begin{bmatrix} (k^2 - \omega^2/c_s^2) & -ik \\ ik & (k^2 c_o^2 q^2/c_s^2 + 1 - \omega^2 Q^2/c_s^2) \end{bmatrix} \begin{Bmatrix} v_o \\ \phi_o \end{Bmatrix} = 0$$

The determinant must be zero for a non-trivial solution hence giving the characteristic equation

$$k^4 - k^2\omega^2\left[\left(\frac{1}{c_s}\right)^2 + \left(\frac{Q}{c_o q}\right)^2\right] - \left[\left(\frac{\omega}{c_o q}\right)^2 - \left(\frac{Q\omega}{c_s c_o q}\right)^4\right] = 0$$

Solving for k gives the spectrum relations as

$$k = \pm\left\{\frac{1}{2}\left[\left(\frac{1}{c_s}\right)^2 + \left(\frac{Q}{c_o q}\right)^2\right]\omega^2 \pm \sqrt{\left(\frac{\omega}{c_o q}\right)^2 + \frac{1}{4}\left[\left(\frac{1}{c_s}\right)^2 - \left(\frac{Q}{c_o q}\right)^2\right]^2\omega^4}\right\}^{\frac{1}{2}} \qquad (4.59)$$

There are four modes just as in the elementary theory, hence the solution structure is also the same as before. The higher-order theory for rods introduced additional modes, but this is not so here. The behavior of each mode, however, is changed as can be seen from Figure 4.18. To explain these plots, consider the following limiting behaviors.

First, consider the limit as the frequency becomes small. Then

$$k_1 \approx \pm\sqrt{\frac{\omega}{c_o q}}, \qquad k_2 \approx \pm i\sqrt{\frac{\omega}{c_o q}}$$

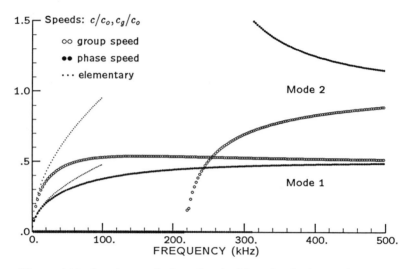

Figure 4.19: Spectrum relations for the Timoshenko beam theory.

These are the Bernoulli-Euler beam spectrum relations. Note that they do not depend on either GAK or ρI and that is why this higher-order theory does not introduce additional modes. On the other hand, as the frequency becomes very large

$$k_1 \Rightarrow \frac{\omega}{c_s}, \qquad k_2 \Rightarrow \frac{Q}{c_o q}\omega$$

Both are propagating solutions, hence the imaginary branch must have turned real. Look for a possible cut-off frequency where k is zero and this is found to be where

$$\omega_0 = \frac{c_s}{Q} = \sqrt{\frac{GAK}{\rho I}} \tag{4.60}$$

To eliminate the rotational inertia, it is necessary to let $Q = 0$ and to eliminate the shear deformation to let $c_s = \infty$. Note that in both cases the cut-off frequency goes to infinity giving the second mode only as evanescent. The limits on the phase and group speeds gives for low frequencies

$$c_g = 2c = \sqrt{c_o q \omega}$$

and for high frequencies

$$c_g = c = c_s \qquad \text{and} \qquad c_g = c = \frac{c_o q}{Q}$$

Both of these high-frequency limits are non-dispersive. The complete behavior is shown in Figure 4.19. The group speed shows a peak before it asymptotes to the same limiting value as the phase speed.

Both the spectrum and dispersion plots show four distinct regions. First, at the low-frequency end, the elementary theory and the Timoshenko show very little difference. At the very-high-frequency limit, the beam behaves non-dispersively with two propagating speeds. In between are two other regions of great practical interest because they show the transition from the elementary theory. Below the cut-off frequency there can be substantial deviation of the two theories, but there is still only one propagating mode. Thus this situation can be described adequately by taking the Timoshenko theory and simplifying it by taking $Q = 0$ or $GAK = \infty$. This simplification may be worthwhile in some circumstances, although it is apparent from this section that, when using the spectral analysis approach, there is hardly any difference in solution complexity between the elementary theory and the full Timoshenko theory.

Just as for the Mindlin-Herrmann rod theory, it is usual to choose a value for K such that the limiting high-frequency behavior is obtained. However, some authors have shown preference for making the cut-off frequency for the second mode as the criterion (see References [42,209] for more discussion). Since the Timoshenko theory is approximate, then neither of these approaches can be judged more right than the other — it is more prudent to choose K so as to give best agreement with the exact solution within the frequency range of interest. Without information to the contrary a value of $K = 0.85$ can be chosen.

Impact of a Timoshenko Beam

The procedure for solving boundary problems for the Timoshenko beam is the same as for the Bernoulli-Euler beam. That is, either displacement or rotation is taken as the basic unknown (the other being obtained from the amplitude ratio obtained from equation (4.58)) and the relations (4.56) used to obtained the resultants. The procedure is illustrated with the problem of the central impact of a Timoshenko beam. This is analogous to the problem already considered section (4.2). A continuous transform solution is given in Reference [53].

The complete solutions for the deformations are

$$\hat{v}(x) = \mathbf{A}e^{-ik_1 x} + \mathbf{B}e^{-ik_2 x} + \mathbf{C}e^{ik_1 x} + \mathbf{D}e^{ik_2 x} \tag{4.61}$$

$$\hat{\phi}(x) = \bar{\mathbf{A}}e^{-ik_1 x} + \bar{\mathbf{B}}e^{-ik_2 x} + \bar{\mathbf{C}}e^{ik_1 x} + \bar{\mathbf{D}}e^{ik_2 x} \tag{4.62}$$

Either can be chosen since the amplitude coefficients are related by

$$\left[GAKk^2 - \rho A\omega^2\right] \text{(unbarred mode)} = [ikGAK] \text{(barred mode)}$$

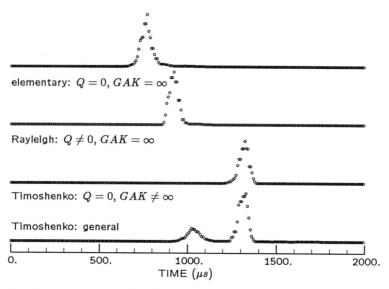

Figure 4.20: Reconstructions showing the presence of a second propagating mode.

For the impact problem, it is simplest to choose the rotation as the working variable. The boundary conditions are then stated in terms of these according to

Displacement: $(\bar{A}, ...) = (A, ...) [ikGAK] / [GAKk^2 - \rho A\omega^2]$

Slope: $\hat{\phi} = \hat{\phi}$

Moment: $\hat{M} = EI\dfrac{d\hat{\phi}}{dx}$

Shear Force: $\hat{V} = -EI\dfrac{d^2\hat{\phi}}{dx^2} - \rho I\omega^2\hat{\phi}$

Choosing the slope on the positive x side to be

$$\hat{\phi}(x) = \bar{A}e^{-ik_1x} + \bar{B}e^{-ik_2x}$$

and imposing the conditions at $x = 0$ of $\hat{\phi} = 0$ and $\hat{V} = -\hat{F}/2$ gives

$$\hat{\phi}(x) = \frac{i\hat{F}}{2EI[k_1^2 - k_2^2]}[e^{-ik_1x} - e^{-ik_2x}] \tag{4.63}$$

Thus, the procedure is essentially the same as used for the elementary theory. The only real difference is that the resultants are slightly more complicated functions of the dependent variable.

To demonstrate the effect of the second propagating mode, a narrow banded force history similar to that of Figure 1.5 was constructed. The peak in the spectrum was made sharp by having many zero crossings and the side lobes were reduced by using a half-sine modulation function. So as to put the central frequency of this pulse (50 kHz) above the cut-off frequency, the beam thickness was increased to 2 inches. (This gave a cut-off of about 35 kHz.) Such an input force generates responses that also have many zero crossings, so the reconstructions of Figure 4.20 show only the amplitude profiles of the reconstructions.

As expected, the elementary theory predicts faster speeds, but more significantly it is lacking the double pulse of the Timoshenko theory. Two intermediate theories are also shown. The Rayleigh theory accounts for rotational inertia but assumes that there is no shear deformation. It predicts a speed similar to one of the Timoshenko pulses. Also shown is what happens when shear deformation is accounted for but the rotational inertia is neglected. The speed predicted by this corresponds to the slower Timoshenko mode. It is worth pointing out that while the phase speed for the second mode goes through an infinity at frequencies near the cut-off, the Timoshenko response does not show any disturbances traveling at these sort of speeds. Therefore, it can be misleading to use the phase speed as an estimate for the propagation of disturbances. This also shows the usefulness of the group speed concept.

Chapter 5

Wave Propagation in Structures

The present chapter is the culmination of the procedures developed thus far: It allows the analysis of complicated connected structures. In the previous chapters, the systems of equations were set up on an *ad hoc* basis. This is reasonable and efficient for simple connectivities but can begin to get unwieldy even in the straightforward example of the elastic joint. The only way to efficiently handle problems with complicated boundaries and discontinuities is to develop a matrix methodology. That is the purpose of this chapter.

An approach similar in style to that of the finite element method of analysis is introduced and has all the advantages of this powerful method. The single most significant difference is that the stiffness matrix is established in the frequency domain, thus allowing the distributed mass to be described exactly. In contrast to the conventional element this means that elements can span all the way from one joint to another and the system of equations to solve are therefore incredibly small. The approach also provides all the other advantages of the spectral analysis method.

To better explain the approach, the first half of this chapter reviews the conventional finite element approach to dynamics problems. The spectral formulation is then discussed in that context.

5.1 Truss and Frame Analysis

A truss consists of a collection of arbitrarily oriented rod members that are slender and cannot support bending loads. The joints are pinned so that they, too, do not support moments. A frame structure, on the other hand, is one that consists of beam members which are connected rigidly at the joints. Thus, the members of a frame must take both bending as well as axial loads, and at the joint the relative positions of the members remain unchanged after deformation. The frame will be concentrated on because the truss can be obtained by reduction from it.

The essential aspect to the study of these structures is to consider the element stiffness of an arbitrarily oriented member. Since differently oriented members are to be considered simultaneously, they then must have a common or global reference frame. More details on the procedures for static structures can be found in Reference [148].

Element Stiffness Matrix for Rods

Consider a rod member of length L and cross-section area A, and denote the end points as node 1 and node 2, respectively, as shown in Figure 5.1. Assume that under the action of external end forces F_1 and F_2, the displacements at the nodes are u_1 and u_2, respectively.

Figure 5.1: Nodal loads and degrees of freedom.

Assume also that the element is small enough so that the distributed mass can be approximated as concentrated masses at the ends. This assumption removes inertia effects from the interior of the element. If there are no other loads applied in between these two nodes, then the equation of motion becomes

$$EA\frac{\partial^2 u}{\partial x^2} = 0$$

This can be integrated directly to give the displacement in the element as a linear function of x:

$$u(x,t) = a_0(t) + x a_1(t)$$

Note that the time dependence enters only through the time dependence of the coefficients. The displacements at each node can be related to the coefficients by imposing that

$$u(0) = u_1 = a_0, \qquad u(L) = u_2 = a_0 + a_1 L$$

This gives

$$a_0 = u_1, \qquad a_1 = -\frac{1}{L}(u_1 - u_2)$$

allowing the displacement distribution to be written in terms of these nodal values as

$$u(x,t) = \left(1 - \frac{x}{L}\right)u_1 + \frac{x}{L}u_2 \equiv f_1(x)u_1(t) + f_2(x)u_2(t) \tag{5.1}$$

Now it is clearer how the time dependence enters through the nodal values. The axial force at an arbitrary position is related to the nodal displacements by

$$F(x) = EA\frac{\partial u}{\partial x} = -\frac{EA}{L}(u_1 - u_2)$$

In this case, the axial force is constant along the length and is directly proportional to the relative nodal displacements. Since the nodal forces are related to the member forces by $F_1 = -F(0)$, $F_2 = F(L)$, then, in matrix notation, this can be expressed in the form

$$\begin{Bmatrix} F_1 \\ F_2 \end{Bmatrix} = \frac{EA}{L}\begin{bmatrix} 1 & -1 \\ -1 & 1 \end{bmatrix}\begin{Bmatrix} u_1 \\ u_2 \end{Bmatrix} \tag{5.2}$$

or symbolically,

$$\{F\} = [\ k\ ]\{u\}$$

in which $[\ k\ ]$ is the *stiffness matrix* for the rod element with respect to the local coordinates.

It is noted that $[\ k\ ]$ in its present form is singular, that is, $\det[\ k\ ] = 0$ and consequently, its inverse does not exist. This implies that given an arbitrary force vector $\{F\}$, it is not possible to find a unique solution for $\{u\}$. A singular stiffness matrix often indicates that the structure is not stable. In the present case, it means that boundary conditions must be specified before there is a solution.

Element Stiffness Matrix for Beams

Consider a straight homogeneous beam element also shown in Figure 5.1. Assume that there are no external loads applied between the two ends at node 1 and node 2. At each node, there are two essential beam actions, namely, the bending moment and shear force. The corresponding nodal degrees of freedom are the rotation $\phi(x, t)$ (or the slope of the deflection curve at the node) and the vertical displacement $v(x, t)$. The positive directions of nodal forces and moments and the corresponding displacements and rotations are shown in the figure.

Again assume that the element is small enough that the inertia properties can be concentrated at the ends. The equation of motion then is

$$EI\frac{\partial^4 v}{\partial x^4} = 0$$

giving the general solution for the deflection curve as

$$v(x, t) = a_0(t) + xa_1(t) + x^2 a_2(t) + x^3 a_3(t)$$

where a_0, a_1, a_2, and a_3 are time varying coefficients. By using the following end conditions

$$v(0) = v_1, \qquad \frac{dv(0)}{dx} = \phi_1$$

$$v(L) = v_2, \qquad \frac{dv(L)}{dx} = \phi_2$$

these coefficients can be solved in terms of the nodal displacements v_1 and v_2 and the nodal rotations ϕ_1 and ϕ_2 as

$$a_0 = v_1, \qquad a_2 = -\frac{3}{L^2}v_1 - \frac{2}{L}\phi_1 + \frac{3}{L^2}v_2 - \frac{1}{L}\phi_2$$

$$a_1 = \phi_1, \qquad a_3 = \frac{2}{L^3}v_1 + \frac{1}{L^2}\phi_1 - \frac{2}{L^3}v_2 + \frac{1}{L^2}\phi_2$$

Substitution of these into the above deflection relation leads to

$$v(x,t) = \left[1 - 3\left(\frac{x}{L}\right)^2 + 2\left(\frac{x}{L}\right)^3\right]v_1 + \left(\frac{x}{L}\right)\left[1 - 2\left(\frac{x}{L}\right) + \left(\frac{x}{L}\right)^2\right]L\phi_1$$

$$+\left(\frac{x}{L}\right)^2\left[3 - 2\left(\frac{x}{L}\right)\right]v_2 + \left(\frac{x}{L}\right)^2\left[-1 + \left(\frac{x}{L}\right)\right]L\phi_2$$

$$\equiv g_1(x)v_1(t) + g_2(x)L\phi_1(t) + g_3(x)v_2(t) + g_4(x)L\phi_2(t) \qquad (5.3)$$

This allows the bending moment and shear force to be determined at an arbitrary position since

$$M(x) = EI\frac{\partial^2 v}{\partial x^2}, \qquad V(x) = -EI\frac{\partial^3 v}{\partial x^3}$$

Further, since these member values and the nodal values are related by

$$m_1 = M(0), \quad V_1 = -V(0), \quad m_2 = M(L), \quad V_2 = V(L)$$

then the matrix form for these relations becomes

$$\left\{\begin{array}{c} V_1 \\ m_1 \\ V_2 \\ m_2 \end{array}\right\} = \frac{EI}{L^3}\left[\begin{array}{cccc} 12 & 6L & -12 & 6L \\ 6L & 4L^2 & -6L & 2L^2 \\ -12 & -6L & 12 & -6L \\ 6L & 2L^2 & -6L & 4L^2 \end{array}\right]\left\{\begin{array}{c} v_1 \\ \phi_1 \\ v_2 \\ \phi_2 \end{array}\right\} \qquad (5.4)$$

Defining the forces and moments and the corresponding generalized displacements as column matrices, the above can be put in the obvious form

$$\{F\} = [\,k\,]\{u\}$$

where $[\,k\,]$ is the beam element stiffness matrix. Note that it is symmetric and is purposely put in a form similar to that for the rod. This will allow the two to be combined.

Global Stiffness Matrix

In a frame or truss consisting of members with different orientations, many local coordinate systems are needed in order that the stiffness matrix as derived in the previous sections can be used. To solve the problem as a whole it is necessary

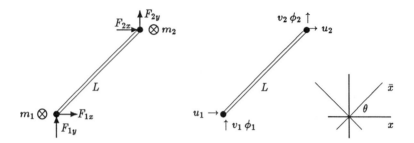

Figure 5.2: Transformation for global axes.

that a single coordinate system is used. For this reason the stiffness matrix of an arbitrarily oriented element will be derived referred to the global coordinate system.

The general member is subjected to both axial and bending loads as indicated in Figure 5.2. If the deflection is small, then flexural deformation and axial deformation are uncoupled; that is, the force-displacement relation can be written separately for each loading type. Consider an element whose longitudinal axis makes an angle θ relative to the x-axis. With respect to the local coordinates $\bar{x}, \bar{y}$, the stiffness matrix is as given for the rod. Augment this equation by adding the trivial forces in the $\bar{y}$-direction and the nodal displacements $\bar{v}_1$ and $\bar{v}_2$ in the $\bar{y}$-direction at node 1 and node 2, respectively. The forces and displacements referred to the local coordinates are now denoted by barred notations. The beam stiffness can be augmented in a similar way. By combining these six equations, the load-deflection relation for the general element can be expressed as

$$\{\bar{F}\} = [\,\bar{k}\,]\{\bar{u}\} \qquad \text{where} \qquad \{\bar{F}\} = \begin{Bmatrix} F_1 \\ V_1 \\ m_1 \\ F_2 \\ V_2 \\ m_2 \end{Bmatrix}, \qquad \{\bar{u}\} = \begin{Bmatrix} u_1 \\ v_1 \\ \phi_1 \\ u_2 \\ v_2 \\ \phi_2 \end{Bmatrix}$$

and

$$[\,\bar{k}\,] = \frac{EA}{L} \begin{bmatrix} 1 & 0 & 0 & -1 & 0 & 0 \\ 0 & 0 & 0 & 0 & 0 & 0 \\ 0 & 0 & 0 & 0 & 0 & 0 \\ -1 & 0 & 0 & 1 & 0 & 0 \\ 0 & 0 & 0 & 0 & 0 & 0 \\ 0 & 0 & 0 & 0 & 0 & 0 \end{bmatrix} + \frac{EI}{L^3} \begin{bmatrix} 0 & 0 & 0 & 0 & 0 & 0 \\ 0 & 12 & 6L & 0 & -12 & 6L \\ 0 & 6L & 4L^2 & 0 & -6L & 2L^2 \\ 0 & 0 & 0 & 0 & 0 & 0 \\ 0 & -12 & -6L & 0 & 12 & -6L \\ 0 & 6L & 2L^2 & 0 & -6L & 4L^2 \end{bmatrix}$$

Thus the stiffness is represented by a 6×6 matrix.

To analyze a general frame it is necessary to know the stiffness matrix of an arbitrary element. Set up the local coordinate system $\bar{x}-\bar{y}$ and the global coordinate system $x-y$ also shown in the figure. The nodal forces referred to these two systems are denoted by

$$\{\bar{F}\} = \begin{Bmatrix} F_1 \\ V_1 \\ m_1 \\ F_2 \\ V_2 \\ m_2 \end{Bmatrix} \quad \text{and} \quad \{F\} = \begin{Bmatrix} F_{x1} \\ F_{y1} \\ m_1 \\ F_{x2} \\ F_{y2} \\ m_2 \end{Bmatrix}$$

respectively. The relation between $\{\bar{F}\}$ and $\{F\}$ is obtained by using the usual coordinate transformation for vectors. Thus

$$\begin{aligned}
F_1 &= \cos\theta\, F_{x1} + \sin\theta\, F_{y1} \\
V_1 &= -\sin\theta\, F_{x1} + \cos\theta\, F_{y1} \\
m_1 &= m_1 \\
F_2 &= \cos\theta\, F_{x2} + \sin\theta\, F_{y2} \\
V_2 &= -\sin\theta\, F_{x2} + \cos\theta\, F_{y2} \\
m_2 &= m_2
\end{aligned}$$

Symbolically, the above equations can be written as

$$\{\bar{F}\} = [\,T\,]\{F\}$$

where $[\,T\,]$ is the transformation or rotation matrix given by

$$[\,T\,] = \begin{bmatrix}
\cos\theta & \sin\theta & 0 & 0 & 0 & 0 \\
-\sin\theta & \cos\theta & 0 & 0 & 0 & 0 \\
0 & 0 & 1 & 0 & 0 & 0 \\
0 & 0 & 0 & \cos\theta & \sin\theta & 0 \\
0 & 0 & 0 & -\sin\theta & \cos\theta & 0 \\
0 & 0 & 0 & 0 & 0 & 1
\end{bmatrix}$$

It can be easily shown that the transformation matrix $[\,T\,]$ is orthogonal, that is, $[\,T\,]^{-1} = [\,T\,]^T$. Define the nodal displacement vectors $\{\bar{u}\}$ and $\{u\}$ as

$$\{\bar{u}\} = \begin{Bmatrix} \bar{u}_1 \\ \bar{v}_1 \\ \phi_1 \\ \bar{u}_2 \\ \bar{v}_2 \\ \phi_2 \end{Bmatrix} \quad \text{and} \quad \{u\} = \begin{Bmatrix} u_1 \\ v_1 \\ \phi_1 \\ u_2 \\ v_2 \\ \phi_2 \end{Bmatrix}$$

respectively. Again, using the coordinate transformation law between the barred and unbarred quantities gives

$$\{\bar{u}\} = [\,T\,]\{u\}$$

Substituting for both $\{\bar{F}\}$ and $\{\bar{u}\}$ into the element stiffness matrix leads to

$$[\,T\,]\{F\} = [\,\bar{k}\,][\,T\,]\{u\}$$

from which the following is obtained

$$\{F\} = [\,T\,]^{-1}[\,\bar{k}\,][\,T\,]\{u\} = [\,T\,]^{T}[\,\bar{k}\,][\,T\,]\{u\} \equiv [\,k\,]\{u\}$$

since $[\,T\,]$ is orthogonal. The explicit expression for the global stiffness matrix $[\,k\,]$ is given as

$$
[\,k\,] = \frac{EA}{L}
\begin{bmatrix}
C^2 & & & & & \text{sym} \\
CS & S^2 & & & & \\
0 & 0 & 0 & & & \\
-C^2 & -CS & 0 & C^2 & & \\
-CS & -S^2 & 0 & CS & S^2 & \\
0 & 0 & 0 & 0 & 0 & 0
\end{bmatrix}
$$

$$
+ \frac{EI}{L^3}
\begin{bmatrix}
12S^2 & & & & & \text{sym} \\
-12CS & 12C^2 & & & & \\
-6LS & 6L & 4L^2 & & & \\
12S^2 & -12CS & 6LS & 12S^2 & & \\
12CS & -12C^2 & -6LC & -12CS & 12C^2 & \\
-6LS & 6LC & 2L^2 & 6LS & -6LC & 4L^2
\end{bmatrix}
\tag{5.5}
$$

where the abbreviations $C \equiv \cos\theta$, $S \equiv \sin\theta$ are used. Note that both of these matrices reduce to the respective matrices when $\theta = 0$.

The essential point to realize is that once the local stiffness relation is obtained, then the global matrix is determined simply by transformation. In this manner, all elements are described in terms of common nodal degrees of freedom.

5.2   Structural Stiffness Matrix

The stiffness property of a frame as a whole can also be characterized by a square, symmetric matrix which relates the external loads with the nodal displacements. There are many ways to establish this and the one chosen here is by imposing nodal equilibrium. This approach is the easiest one to extend to dynamic problems.

Assemblage

To illustrate the procedure for constructing the structural stiffness matrix, consider a node connecting three members as shown in the Figure 5.3. Assume that the structure is in a state of equilibrium; thus each node (joint) must also be in a state of equilibrium.

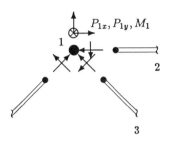

Figure 5.3: Resultant loads at a node.

A free body diagram containing node 1 under the action of externally applied forces and moments (P_x, P_y, M) are shown in the figure. The nodal forces in member 1-2 referred to the global coordinates are denoted by F_{x1}^{12} and F_{y1}^{12}. Note that the action of the member on the joint is equal and opposite. The equilibrium equations are

$$\begin{aligned} P_{x1} &= F_{x1}^{12} + F_{x1}^{13} + \cdots \\ P_{y1} &= F_{y1}^{12} + F_{y1}^{13} + \cdots \\ M_1 &= m_1^{12} + m_1^{13} + \cdots \end{aligned}$$

Equilibrium equations similar to these can be set up for all the other nodes. The combined equations can then be expressed in the form

$$\begin{Bmatrix} P_{x1} \\ P_{y1} \\ M_1 \\ \vdots \\ M_n \end{Bmatrix} = \begin{Bmatrix} F_{x1} \\ F_{y1} \\ m_1 \\ \vdots \\ 0 \end{Bmatrix}^{(12)} + \begin{Bmatrix} F_{x1} \\ F_{y1} \\ m_1 \\ \vdots \\ 0 \end{Bmatrix}^{(13)} + \cdots$$

where there is a force vector for each member. The vectors on the right-hand side are related to the nodal displacements $\{u_1, \cdots, v_3\}$ through the 6×6 global element stiffness matrices $[k^{(12)}]$, $[k^{(23)}]$ and $[k^{(13)}]$ by augmenting to a $N \times N$ matrix. For instance,

$$\begin{Bmatrix} F \\ 0 \end{Bmatrix}^{(12)} = \begin{bmatrix} [k^{(12)}] & 0 \\ 0 & 0 \end{bmatrix} \{u\}$$

Then all can be combined to give

$$\{P\} = [K^{(12)}]\{u\} + [K^{(13)}]\{u\} + \cdots$$

In shorthand this can be written as

$$\{P\} = [K]\{u\} \qquad \text{or} \qquad P_i = \sum_j K_{ij}u_j$$

It is worth noting that the matrix entry K_{ij}, which relates P_i with u_j, is the sum of all the entries in the element stiffness matrices that relate P_i and u_j. Thus the stiffness property of a structure as a whole is characterized by a symmetric square matrix $[K]$ which relates the external loads with the nodal displacements.

Also note that since the individual augmented stiffnesses have many zeroes, then the resultant matrix gives a banded effect. Further, although each element has different forces, they all share common displacements at the nodal points. This automatically ensures that continuity of displacement from element to element. A more detailed description of this assembling procedure can be found in Reference [233].

Boundary Conditions

External forces or displacements may be prescribed at the nodes of the structural system. However, when the force is given, then the corresponding displacement must become an unknown quantity. On the other hand, if the displacement is known, then the corresponding external forces are part of the solution sought. The global system of equations must be rearranged and renumbered to account for this.

To fix ideas, consider a case of two nodes (with six degrees of freedom) where the nodal values are renumbered so that the arrays can be partitioned into known and unknown parts as

$$\{u\} = \left\{ \begin{array}{c} u_u \\ - \\ u_k \end{array} \right\}, \qquad \{P\} = \left\{ \begin{array}{c} P_k \\ - \\ P_u \end{array} \right\}$$

where

$$\{u_u\} = \left\{ \begin{array}{c} u_1 \\ u_2 \end{array} \right\}, \quad \{u_k\} = \left\{ \begin{array}{c} u_3 \\ u_4 \\ u_5 \\ u_6 \end{array} \right\} \qquad \text{and} \qquad \{P_k\} = \left\{ \begin{array}{c} P_1 \\ P_2 \end{array} \right\}, \quad \{P_u\} = \left\{ \begin{array}{c} P_3 \\ P_4 \\ P_5 \\ P_6 \end{array} \right\}$$

The equations for the whole structural system are expressed in the partitioned form as

$$\left\{ \begin{array}{c} P_k \\ P_u \end{array} \right\} = \left[\begin{array}{cc} K_{aa} & K_{ab} \\ K_{ba} & K_{bb} \end{array} \right] \left\{ \begin{array}{c} u_u \\ u_k \end{array} \right\}$$

Multiplying this out gives

$$\{P_k\} = [K_{aa}]\{u_u\} + [K_{ab}]\{u_k\}$$
$$\{P_u\} = [K_{ba}]\{u_u\} + [K_{bb}]\{u_k\} \tag{5.6}$$

The unknown displacements can therefore be obtained from the first equation as

$$[K_{aa}]\{u_u\} = \{P_k\} - \{\Delta\}, \qquad \{\Delta\} = [K_{ab}]\{u_k\} \tag{5.7}$$

which has only known quantities on the right-hand side. The unknown external loads are obtained from a subsequent computation using the second equation above

$$\{P_u\} = [K_{ba}][K_{aa}]^{-1}(\{P_k\} - \{\Delta\}) + [K_{bb}]\{u_k\}$$

Again, everything on the right-hand side is known. This method of rearranging the equations reduces the size of the system to be solved. That is, $[K_{aa}]$ is smaller than $[K]$ and this sometimes can make a substantial difference in the number of computations.

If the known nodal displacements are zero (such as at a fixed boundary), then $\{u_k\} = 0$ and the above take on the particularly simple form

$$[K_{aa}]\{u_u\} = \{P_k\}, \qquad \{P_u\} = [K_{ba}]\{u_u\} \tag{5.8}$$

In this case $[K_{aa}]$ is called the *reduced structural stiffness* matrix relating the unknown displacements to the given external loads.

Computation of Member Loads

Once the nodal displacements are obtained, the nodal forces and moments for each member can be calculated from

$$\{F\} = [\,k\,]\{u\}$$

Using the transformation law, then

$$\{\bar{F}\} = [\,T\,]\{F\} = [\,T\,][\,k\,]\{u\}$$

The axial force F, the shear force V, and the bending moment M throughout the frame member can be obtained from the nodal components. Using the strength of materials sign convention, these are given by

$$F(x) = -\bar{F}_1 = \bar{F}_2$$
$$V(x) = -\bar{V}_1 = \bar{V}_2$$
$$M(x) = -\bar{m}_1 + \frac{x}{L}(\bar{m}_1 + \bar{m}_2) \tag{5.9}$$

The distribution of deflection and slope is best obtained by utilizing the shape functions and the nodal values. That is, they are given as

$$
\begin{aligned}
u(x) &= f_1(x)u_1 + f_2(x)u_2 \\
v(x) &= g_1(x)v_1 + g_2(x)L\phi_1 + g_3(x)v_2 + g_4(x)L\phi_2 \\
\phi(x) &= g_1'(x)v_1 + g_2'(x)L\phi_1 + g_3'(x)v_2 + g_4'(x)L\phi_2
\end{aligned} \tag{5.10}
$$

where the prime refers to differentiation with respect to x.

Computer Implementation

The algorithms for implementing this on a computer are now quite standard and found in many textbooks. The flow is to have a big do-loop over all the elements. For each element, the local $[6 \times 6]$ element stiffness is established and then is transformed to global coordinates. This is entered into the structural stiffness matrix simply by associating the appropriate nodal numbers. When the do-loop is complete, the system is reduced by removing rows and columns associated with the zero degree of freedom boundary conditions. The non-zero nodal degrees of freedom of the resulting system are solved for using one of the standard system solvers that takes advantage of the banded symmetric form of the matrix.

Various efficiencies and enhancements can be introduced but the above is the basic scheme used. A significant point is that many of the procedures are quite standard and easily available.

5.3 Matrix Formulation of Inertia Effects

The wave motion in a structure occurs because of the inertia terms in the equations of motion. The element, as formulated thus far, does not have any mass distribution and therefore cannot exhibit wave motion. The situation is remedied, somewhat, by considering the members themselves (of the structure) to be made of many small elements. Thus the concentrated masses at the end of each element effectively approximate the distributed mass along the member. The price is that substantially more elements must be used than otherwise would be necessary for a static analysis.

The only difference in the dynamic formulation is the effect of the distributed mass. For convenience, this will be assumed to act as an additional force (i.e., the inertia force) acting at each node, and the problem can then be assembled as for the static problem.

Mass Matrix for Rods

The simplest way of evaluating the equivalent inertia forces is through the lump-mass method. That is, replace the distributed mass of the rod by a concentrated

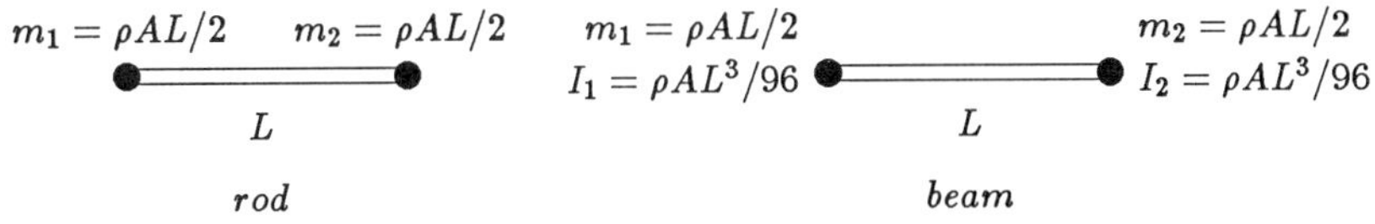

Figure 5.4: Lumped-mass decomposition.

one, and then simply put half of it at node 1 and the other half at 2. If the equivalent inertia forces are denoted by F_1'' and F_2'', then

$$\text{Inertia at 1:}\quad F_1'' \;=\; m\ddot{u}_1 \;=\; \frac{\rho AL}{2}\ddot{u}_1$$

$$\text{Inertia at 2:}\quad F_2'' \;=\; m\ddot{u}_2 \;=\; \frac{\rho AL}{2}\ddot{u}_2$$

or in matrix form

$$\left\{ \begin{matrix} F_1'' \\ F_2'' \end{matrix} \right\} = \frac{\rho AL}{2} \begin{bmatrix} 1 & 0 \\ 0 & 1 \end{bmatrix} \left\{ \begin{matrix} \ddot{u}_1 \\ \ddot{u}_2 \end{matrix} \right\}$$

This suggests the shorthand

$$\{F''\} = [\,m\,]\{\ddot{u}\} \qquad \text{with} \qquad [\,m\,] \equiv \frac{\rho AL}{2} \begin{bmatrix} 1 & 0 \\ 0 & 1 \end{bmatrix}$$

The matrix $[\,m\,]$ is called the mass matrix. It is always symmetric and, in this case, is also diagonal.

An alternative approach to estimating the effect of the mass is to establish an equivalence between the energies of the original and the equivalent system. The kinetic energy of a general system of connected masses are related to the masses and velocities by

$$T = \tfrac{1}{2}\sum_i \sum_j m_{ij}\dot{u}_i\dot{u}_j, \quad \text{or} \quad m_{ij} = \frac{\partial^2 T}{\partial \dot{u}_i \partial \dot{u}_j} \tag{5.11}$$

The second form gives a simple scheme for obtaining the masses from the kinetic energy. The actual kinetic energy is obtained using the functional form of the displacement and is

$$\begin{aligned} T &= \tfrac{1}{2}\int_0^L \rho A[\dot{u}]^2\,dx = \tfrac{1}{2}\rho A \int_0^L [\dot{u}(1 - \tfrac{x}{L}) + \tfrac{x}{L}\dot{u}_2]^2\,dx \\ &= \tfrac{1}{2}\rho AL\{\tfrac{1}{3}\dot{u}_1^2 + \tfrac{1}{3}\dot{u}_1\dot{u}_2 + \tfrac{1}{3}\dot{u}_2^2\} \end{aligned}$$

Using this in the above relation gives the equivalent mass matrix by differentiation as

$$[\,m\,] \equiv \frac{\rho AL}{6} \begin{bmatrix} 2 & 1 \\ 1 & 2 \end{bmatrix} \tag{5.12}$$

The equivalent inertia forces can therefore be written in matrix form as

$$\left\{ \begin{array}{c} F_1'' \\ F_2'' \end{array} \right\} = \frac{\rho A L}{6} \begin{bmatrix} 2 & 1 \\ 1 & 2 \end{bmatrix} \left\{ \begin{array}{c} \ddot{u}_1 \\ \ddot{u}_2 \end{array} \right\} \tag{5.13}$$

Note that these masses do not necessarily have any simple interpretation of masses at nodes. Also, the matrix is no longer diagonal but still symmetric.

Mass Matrix for Beams

To obtain the lumped-mass matrix for a beam, replace the distributed mass by concentrated ones at nodes 1 and 2. Since there is also rotational inertia, this too must be distributed. Note that the pivoting of the beam near its center has the effect of reducing the overall rotational inertia. If the equivalent inertia forces and moments are denoted by F_1'', F_2'' and m_1'', m_2'', respectively, then

Inertia at 1: $\quad F_1'' = m\ddot{v}_1 = \dfrac{\rho A L}{2}\ddot{v}_1\,, \qquad m_1'' = I\ddot{\phi}_1 = \dfrac{\rho A L^3}{96}\ddot{\phi}_1$

Inertia at 2: $\quad F_2'' = m\ddot{v}_2 = \dfrac{\rho A L}{2}\ddot{v}_2\,, \qquad m_2'' = I\ddot{\phi}_2 = \dfrac{\rho A L^3}{96}\ddot{\phi}_2$

or in matrix form

$$\left\{ \begin{array}{c} V_1'' \\ m_1'' \\ V_2'' \\ m_2'' \end{array} \right\} = \frac{\rho A L}{96} \begin{bmatrix} 48 & 0 & 0 & 0 \\ 0 & L^2 & 0 & 0 \\ 0 & 0 & 48 & 0 \\ 0 & 0 & 0 & L^2 \end{bmatrix} \left\{ \begin{array}{c} \ddot{v}_1 \\ \ddot{\phi}_1 \\ \ddot{v}_2 \\ \ddot{\phi}_2 \end{array} \right\}$$

Again this suggests the representation

$$\{F''\} = [\,m\,]\{\ddot{u}\}$$

where $[\,m\,]$ is called the mass matrix. It is symmetric and (in this case) diagonal.

A consistent mass matrix can also be set up by establishing an equivalence between the energies of the original and the equivalent systems as done for the rod. The actual kinetic energy can be obtained using the functional form of the displacement, as

$$\begin{aligned} T &= \tfrac{1}{2}\int_0^L \rho A[\,\dot{v}\,]^2\, dx \\ &= \tfrac{1}{2}\rho A \int_0^L [\dot{v}_1 + \frac{x^2}{L^3}(3L - 2x)(\dot{v}_2 - \dot{v}_1) - x(1 - \frac{x}{L})\dot{\phi}_1 \\ &\qquad\qquad + \frac{x^2}{L^2}(L - x)(\dot{\phi}_1 + \dot{\phi}_2)]^2\, dx \end{aligned}$$

On multiplying out and integrating, this gives the kinetic energy as

$$T = \tfrac{1}{2}\rho A\{\tfrac{1}{105}L^3(\dot\phi_2 - \dot\phi_1)^2 + \tfrac{13}{35}L(\dot v_1 - \dot v_2)^2 + \tfrac{1}{210}L^3\dot\phi_1\dot\phi_2 + \tfrac{1}{6}L^2(\dot\phi_2 - \dot\phi_1)\dot v_1$$
$$- \tfrac{13}{210}L^2\dot\phi_1(\dot v_2 - \dot v_1) + \tfrac{11}{105}L^2\dot\phi_2(\dot v_2 - \dot v_1) + L\dot v_1\dot v_2\}$$

The elements of the mass matrix are now obtained by differentiation and gives in matrix notation (after some rearranging)

$$\left\{\begin{array}{c} V_1'' \\ m_1'' \\ V_2'' \\ m_2'' \end{array}\right\} = \frac{\rho AL}{420} \begin{bmatrix} 156 & 22L & 54 & -13L \\ 22L & 4L^2 & 13L & -3L^2 \\ 54 & 13L & 156 & -22L \\ -13L & -3L^2 & -22L & 4L^2 \end{bmatrix} \left\{\begin{array}{c} \ddot v_1 \\ \ddot\phi_1 \\ \ddot v_2 \\ \ddot\phi_2 \end{array}\right\} \tag{5.14}$$

or simply

$$\{F''\} = [\, m \,]\{\ddot u\}$$

Note that again these masses do not necessarily have any simple interpretation of masses at nodes. Looking at the diagonal terms in comparison with the lumped system, it is seen that

$$\frac{13}{35} \quad vs. \quad \frac{1}{2}, \qquad \frac{1}{105} \quad vs. \quad \frac{1}{96}$$

Thus, the lumped system appears to overestimate the inertias on the diagonal, and (because of the zeros) underestimate the off-diagonal terms. The result is about the same.

Assemblage

The approach to assemblage is similar to that for the static stiffness except that now the element nodal forces are comprised of both elastic and inertia forces as

$$\{F\} = \{F'\} + \{F''\} = [\, k \,]\{u\} + [\, m \,]\{\ddot u\}$$

Dynamic equilibrium is then considered at each node to give, for instance,

$$P_{1x} - F_{1x}^{(12)} - F_{1x}^{(13)} - \cdots = m_1\ddot u_1$$

where m_1 may be a concentrated nodal inertia (separate from that of redistributing the element mass). Substituting for the respective force vectors in terms of the stiffness and mass matrices gives after assembling

$$[\, K \,]\{u\} + [\, M \,]\{\ddot u\} = \{P\}$$

where $[\, K \,]$ is the structural stiffness matrix, while $[\, M \,]$ is the structural mass matrix which also contains any concentrated inertias. It is important to note that the structural mass matrix is assembled in the same way as the stiffness matrix.

In summary, for dynamic problems the assembled stiffness matrix is exactly the same as for the static problem, while the mass matrix is assembled in a completely identical manner. As a result, the mass matrix will exhibit all the symmetry properties of the stiffness matrix. This, perhaps surprising result, will become obvious when it is shown in the next section that the above can be viewed as a two-term expansion of the exactly formulated dynamic element.

Dynamic Problems

The above equations are for the general case of the forced motion of a system of frame members. They are to be interpreted as a system of differential equations in time for the unknown nodal displacements $\{u\}$, subject to the known forcing histories $\{P\}$. Generally, these require some numerical scheme for integration over time. Aspects of this are discussed in Reference [185].

Thus the conventional element approach to dynamics problems pays a price in two respects. First, a substantial increase in the number of elements must be used in order to accurately model the mass distribution. This is true even if the structure itself is simple, for example, just a rod. The other is that the complete system of equations must be solved at each time increment and because of stability requirements, each time step must be quite small. The combination of these two factors has made finite element analysis of wave propagation problems very expensive and feasible only on large mainframe computers.

For the special case when the excitation force is harmonic, that is,

$$\{P\} = \{\hat{P}\}\, e^{i\omega t} \qquad \text{or} \qquad \begin{Bmatrix} P_1 \\ P_2 \\ \vdots \\ P_n \end{Bmatrix} = \begin{Bmatrix} \hat{P}_1 \\ \hat{P}_2 \\ \vdots \\ \hat{P}_n \end{Bmatrix} e^{i\omega t}$$

then the response is also harmonic, given by

$$\{u\} = \{\hat{u}\}\, e^{i\omega t} \qquad \text{or} \qquad \begin{Bmatrix} u_1 \\ u_2 \\ \vdots \\ u_n \end{Bmatrix} = \begin{Bmatrix} \hat{u}_1 \\ \hat{u}_2 \\ \vdots \\ \hat{u}_n \end{Bmatrix} e^{i\omega t}$$

This type of analysis is referred to as forced frequency response. (Note that many of the P_n could be zero.) Substituting these forms into the differential equations gives

$$[K]\{\hat{u}\}e^{i\omega t} - \omega^2[M]\{\hat{u}\}e^{i\omega t} = \{\hat{P}\}e^{i\omega t}$$

or after canceling through the common time factor

$$\left[[K] - \omega^2[M]\right]\{\hat{u}\} = \{\hat{P}\} \qquad \text{or} \qquad [\hat{K}]\{\hat{u}\} = \{\hat{P}\} \qquad (5.15)$$

Thus, the solution can be obtained analogous to the static problem, where the stiffness matrix is modified by the kinetic energy term $\omega^2[M]$ and is therefore frequency dependent. This system of equations is now recognized as the spectral form of the equations of motion of the structure. One approach, then, to wave propagation problems is to evaluate the above at each frequency and then use the FFT for time domain reconstructions. This is feasible but a more full fledged spectral approach is developed in the next few sections.

Before leaving this section, a word about a case of very special interest, namely that of free vibrations. This case gives the undamped mode shapes that are very important in a modal analysis. For free vibrations of the system, the applied loads are zero, i.e., $\{P\} = 0$, giving the equations of motion as

$$\left[[K] - \omega^2[M]\right]\{\hat{u}\} = 0$$

This is a system of homogeneous equations for the nodal displacements $\{\hat{u}\}$. For a nontrivial solution, the determinant of the matrix of coefficients must be zero, i.e.,

$$\det\left|[K] - \omega^2[M]\right| = 0$$

This equation can be turned into the standard form for an eigenvalue problem by multiplying both sides by $[M]^{-1}$, i.e.,

$$\det\left|[M]^{-1}[K] - \omega^2[I]\right| = 0$$

The ω^2 are then the eigenvalues and the corresponding $\{\hat{u}\}$ the eigenvectors of the problem. A vanishing determinant thus indicates resonances. Further, the larger the number of elements, the larger the system, and thus the higher the order of the characteristic equation. Consequently, there are more roots and more resonances. It is important to realize that the determinant can be zero quite independently of the applied loads.

5.4   Spectral Element for Rods

The material developed in Chapter 3 is now used to derive a spectrally formulated finite element. The major significance of this element is that it treats the mass distribution exactly and therefore wave propagation within each element is treated exactly. It also means that the subdivision of the member into many small elements is no longer necessary. More details can be found in Reference [69].

Dynamic Stiffness

The spectral formulation begins with the equations of motion of the rod without neglecting inertia. The element looks the same as in Figure 5.1 except that the

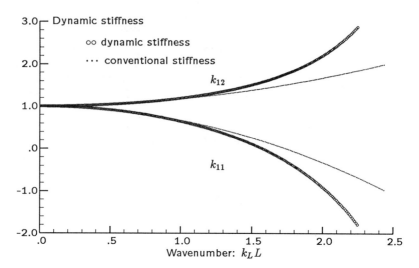

Figure 5.5: Comparison of conventional and spectral element stiffnesses.

mass is no longer concentrated at the ends. It does, however, assume that there are
no applied loads between the rod ends.

It has already been established that the general solution for the rod can be
represented by

$$u(x,t) = \sum_n \hat{u}_n(x, \omega_n) e^{i\omega_n t}$$

where the spectral displacements $\hat{u}_n$ have the simple solution

$$\hat{u}_n(x) = \mathbf{A}e^{-ik_n x} + \mathbf{B}e^{-ik_n(L-x)}, \qquad k_n = \omega_n \sqrt{\frac{\rho A}{EA}}$$

In finite element terms, this is called the *shape function*, but obviously in this case
it is dependent on the frequency ω_n. That is, it is different at each frequency unlike
the shape function of equation (5.1). The coefficients can be related to the nodal
displacements as

$$\left\{ \begin{array}{c} \mathbf{A} \\ \mathbf{B} \end{array} \right\} = \frac{1}{(1 - e^{-i2kL})} \left[\begin{array}{cc} 1 & -e^{-ikL} \\ -e^{-ikL} & 1 \end{array} \right] \left\{ \begin{array}{c} \hat{u}_1 \\ \hat{u}_2 \end{array} \right\}$$

The nodal forces are determined by differentiation of the displacements and when
expressed in matrix form are

$$\left\{ \begin{array}{c} \hat{F}_1 \\ \hat{F}_2 \end{array} \right\} = \frac{EA}{L} \frac{ikL}{(1 - e^{-i2kL})} \left[\begin{array}{cc} 1 + e^{-i2kL} & -2e^{-ikL} \\ -2e^{-ikL} & 1 + e^{-i2kL} \end{array} \right] \left\{ \begin{array}{c} \hat{u}_1 \\ \hat{u}_2 \end{array} \right\} \qquad (5.16)$$

This can be written in the now familiar form of $\hat{F} = [\ \hat{k}\ ]\{\hat{u}\}$ where $[\ \hat{k}\ ]$ is the frequency dependent dynamic element stiffness for the rod. It is symmetric and, surprisingly, it is also real. This is confirmed by expanding the above to

$$\left\{ \begin{array}{c} \hat{F}_1 \\ \hat{F}_2 \end{array} \right\} = \frac{EA}{L}\frac{kL}{\sin kL} \left[\begin{array}{cc} \cos kL & -1 \\ -1 & \cos kL \end{array} \right] \left\{ \begin{array}{c} \hat{u}_1 \\ \hat{u}_2 \end{array} \right\} \tag{5.17}$$

Actually, the exponential form is preferable to program because it allows the introduction of damping into the system and as a result offers more numerical stability.

It must be emphasized that the above is not the spectral form of equation (5.15) because the two element stiffnesses are fundamentally different. To elaborate on this, look at Figure 5.5 which compares the dynamic stiffnesses at small frequencies. For the conventional element using the equivalent mass matrix, this is

$$\left\{ \begin{array}{c} \hat{F}_1 \\ \hat{F}_2 \end{array} \right\} = \frac{EA}{L} \left[\begin{array}{cc} (1 - \frac{1}{3}\xi^2) & -(1 + \frac{1}{6}\xi^2) \\ -(1 + \frac{1}{6}\xi^2) & (1 - \frac{1}{3}\xi^2) \end{array} \right] \left\{ \begin{array}{c} \hat{u}_1 \\ \hat{u}_2 \end{array} \right\}, \qquad \xi \equiv \omega\sqrt{\frac{\rho A}{EI}} = kL$$

Thus the conventional stiffness is monotonic in frequency and for $\hat{k}_{11}$, for example, goes through zero only once. The spectral dynamic stiffness, on the other hand, has many zeros as seen from Figure 5.6. A stiffness that has such wild behavior may seem strange but consider the following: The above conventional stiffness is for a single element of length L. Now replace it by two elements each of length $\frac{1}{2}L$ and assemble them. The resulting stiffness relation is

$$\left\{ \begin{array}{c} \hat{F}_1 \\ 0 \\ \hat{F}_2 \end{array} \right\} = \frac{2EA}{L} \left[\begin{array}{ccc} (1 - \frac{1}{12}\xi^2) & -(1 + \frac{1}{24}\xi^2) & 0 \\ -(1 + \frac{1}{24}\xi^2) & (1 - \frac{1}{12}\xi^2) & -(1 + \frac{1}{24}\xi^2) \\ 0 & -(1 + \frac{1}{24}\xi^2) & (1 - \frac{1}{12}\xi^2) \end{array} \right] \left\{ \begin{array}{c} \hat{u}_1 \\ \hat{u}^* \\ \hat{u}_2 \end{array} \right\}$$

The middle force is zero because there are no applied loads there. Now solve for $\hat{u}^*$ in terms of the other displacements and remove it from the system to give

$$\left\{ \begin{array}{c} \hat{F}_1 \\ \hat{F}_2 \end{array} \right\} = \frac{2EA}{L}\frac{1}{2(1 - \frac{1}{12}\xi^2)} \left[\begin{array}{cc} (1 - \frac{5}{12}\xi^2 + \frac{1}{192}\xi^4) & -(1 + \frac{1}{24}\xi^2)^2 \\ -(1 + \frac{1}{24}\xi^2)^2 & (1 - \frac{5}{12}\xi^2 + \frac{1}{192}\xi^4) \end{array} \right] \left\{ \begin{array}{c} \hat{u}_1 \\ \hat{u}_2 \end{array} \right\}$$

The behavior of $\hat{k}_{11}$, for instance, is that it crosses zero near $kL = 1.57$ compared to 1.73 before. This is closer to the spectral value of $\pi/2 = 1.57$. What is more important, however, is that it has a second zero and it also goes through an infinity at $kL = \sqrt{12} = 3.46$. Thus the apparently odd behavior of Figure 5.6 is also implied in the conventional formulation — if a sufficient number of elements are used. The spectral approach is equivalent to an infinite number of conventional elements.

The significant difference, from a practical point of view, between the two formulations is that the number of rod segments in the present formulation need only coincide with the number of discontinuities. Thus only one element need be used per uniform segment. This can result in an enormous reduction in the size of the matrices to be solved. By now, also, it is obvious that the spectral approach has all the other advantages described in the previous chapters.

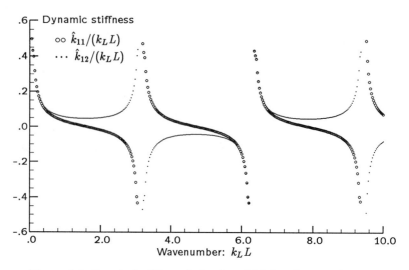

Figure 5.6: Spectral stiffness behavior at higher frequencies.

Throw-off Bar

The boundary conditions to be imposed are quite similar to those of the conventional element and they are imposed in the usual way. There is, however, an additional condition that is unique to wave problems. It corresponds to when the element extends to infinity, that is, the element behaves as a throw-off bar in that it is a conduit for energy out of the system. Its form simply is

$$\hat{F}_1 = EA[ik]\hat{u}_1 \tag{5.18}$$

Note that it is complex and therefore the structural stiffness matrix is also complex after assembling. This boundary condition is very useful when the time of interest is short and the structure is large. Then the structure can be partitioned into local and remote parts, and the connections replaced with throw-off bars that conduct energy out of the local substructure. The primary effect of this is to allow fine time resolution for a given spectral size N. This element also acts effectively as a damper to the system and thus improves the quality of the numerical calculations.

Responses at Arbitrary Locations

Since the spectral element can be very large, then the ability to compute the response between nodes is necessary. This can be done simply by using the shape function and the nodal values. This is akin to the problem of finding member loads in the conventional formulation. Thus, the displacements and forces are obtained from

$$\hat{u}(x) = \left(\frac{\sin k(L-x)}{\sin kL}\right)\hat{u}_1 + \left(\frac{\sin kx}{\sin kL}\right)\hat{u}_2 \tag{5.19}$$

$$\hat{F}(x) = -EAk\left(\frac{\cos k(L-x)}{\sin kL}\right)\hat{u}_1 + EAk\left(\frac{\cos kx}{\sin kL}\right)\hat{u}_2 \tag{5.20}$$

The time reconstructions, of course, are obtained by using the FFT. Note that these are used after the nodal values have been obtained (i.e., as part of the post-processing) and need only be applied to the members of interest.

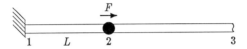

Figure 5.7: Two element semi-infinite rod.

A Simple Example

Consider the problem of a fixed rod shown in Figure 5.7 and impacted with an inline force. This is modeled as two elements where the second is considered as extending to infinity.

   The reduced stiffness matrix gives non-zero behavior only at node 2, and is

$$\{\hat{F}_2\} = [\frac{EA}{L}kL\frac{\cos kL}{\sin kL} + EAik]\{\hat{u}_2\} = \frac{EAk}{\sin kL}[e^{ikL}]\{\hat{u}_2\}$$

Hence the response is, on replacing $\sin kL$ by its exponential form,

$$\{\hat{u}_2\} = \frac{1}{i2EAk}[1 - e^{-i2kL}]\{\hat{F}_2\}$$

This represents two pulses shifted in time by the amount needed to travel a distance $2L$. The magnitude of the pulse in one-half that of the input. This is shown in Figure 5.8. The meaning of the figure is that the initial pulse causes a forward-moving compressive wave and a backward-moving tensile one. This latter pulse, on reflection from the fixed end, retains its sign and so is picked up later as a compressive pulse at the impacted node. What is particularly significant about this example is that although the basic element stiffness exhibits singularities, the resultant structural stiffness does not in this case.

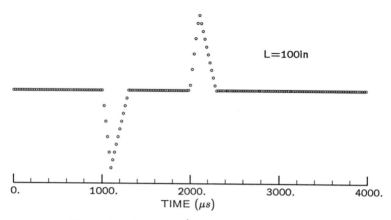

L=100in

Figure 5.8: Response near the impact site.

5.5   Spectral Element for Beams

The approach here follows much the same as the previous section but since (as shown in Chapter 4) beams are two-mode systems, then the stiffness matrix is of order 4×4 because there must be two degrees of freedom at each node (see Reference [75] for more details). An interesting application of this approach to split beams can be found in Reference [73].

Dynamic Stiffness

The spectral formulation begins with the equation of motion of the beam including the inertia term. It looks similar to that of Figure 5.1 where again it is assumed that there are no applied loads between the element ends. Just as for the rod, the displacements are given their spectral representation

$$v(x,t) = \sum_n \hat{v}_n(x,\omega_n)e^{i\omega_n t}$$

The spectral components, $\hat{v}_n$, have been shown to have the simple solution

$$\hat{v}(x) = \mathbf{A}e^{-ik_n x} + \mathbf{B}e^{-k_n x} + \mathbf{C}e^{-ik_n(L-x)} + \mathbf{D}e^{-k_n(L-x)}, \qquad k_n = \sqrt{\omega_n}\left[\frac{\rho A}{EI}\right]^{1/4}$$

where the first two are appropriate to waves moving in the plus direction and the second two to backward-moving waves. Both sets are necessary since the element is finite.

The dynamic stiffness is set up as before by first relating the coefficients to the nodal displacements as

$$\{\mathbf{A},\mathbf{B},\mathbf{C},\mathbf{D}\} = [\,\hat{G}\,]\{\hat{v}_1,\hat{\phi}_1,\hat{v}_2,\hat{\phi}_2\} \tag{5.21}$$

The matrix $[\hat{G}]$ is fairly complicated so it is not repeated here. The nodal loads are obtained by differentiation and can be rearranged to give

$$\{\hat{F}\} = \frac{EI}{L^3}[\ \hat{k}\ ]\{\hat{u}\}$$

The individual stiffness terms are a little more complicated than for the rod giving, for instance,

$$
\begin{aligned}
\hat{k}_{11} &= i(1+i)(z_{11}z_{22} + iz_{12}z_{21})/(z_{11}^2 + z_{12}^2)\\
\hat{k}_{22} &= i(1+i)(z_{11}z_{22} - iz_{12}z_{21})/(z_{11}^2 + z_{12}^2)
\end{aligned}
\tag{5.22}
$$

where

$$z_{11} = 1 - e^{-ikL}e^{-kL}, \quad z_{22} = 1 + e^{-ikL}e^{-kL}, \quad z_{12} = e^{-ikL} - e^{-kL}, \quad z_{21} = e^{-ikL} + e^{-kL}$$

The matrix is real and symmetric as can be seen from its reduced form

$$
\begin{Bmatrix} \hat{V}_1 \\ \hat{m}_1 \\ \hat{V}_2 \\ \hat{m}_2 \end{Bmatrix}
= \frac{EI}{L^3}
\begin{bmatrix}
\alpha & \bar{\gamma}L & -\bar{\alpha} & \gamma L \\
 & \beta L^2 & -\gamma L & \bar{\beta}L^2 \\
 & & \alpha & -\bar{\gamma}L \\
\text{sym} & & & \beta L^2
\end{bmatrix}
\begin{Bmatrix} \hat{v}_1 \\ \hat{\phi}_1 \\ \hat{v}_2 \\ \hat{\phi}_2 \end{Bmatrix}
\tag{5.23}
$$

where

$$
\begin{aligned}
\alpha &= (CSh + SCh)(kL)^3/\det, & \bar{\alpha} &= (S + Sh)(kL)^3/\det\\
\beta &= (-CSh + SCh)(kL)/\det, & \bar{\beta} &= (-S + Sh)(kL)/\det\\
\gamma &= (-C + Ch)(kL)^2/\det, & \bar{\gamma} &= (SSh)(kL)^2/\det
\end{aligned}
$$

$$\det = 1 - CCh, \ C \equiv \cos kL, \ S \equiv \sin kL, \ Ch \equiv \cosh kL, \ Sh \equiv \sinh kL$$

Reference [35] introduced a similar matrix to study vibration problems. Again, for programming purposes the exponential form is preferable.

For comparison, the form for the conventional element with the consistent mass matrix can be written as above, but with the associations

$$
\begin{aligned}
\alpha &= 12 - (kL)^4\frac{13}{35}, & \bar{\alpha} &= 12 + (kL)^4\frac{9}{70}\\[2mm]
\beta &= 4 - (kL)^4\frac{1}{105}, & \bar{\beta} &= 2 + (kL)^4\frac{1}{140}\\[2mm]
\gamma &= 6 + (kL)^4\frac{13}{420}, & \bar{\gamma} &= 6 - (kL)^4\frac{11}{210}
\end{aligned}
$$

(This assumes harmonic excitation.) Figure 5.9 shows a comparison between the present and conventional dynamic stiffnesses. It is apparent that they both have the same limiting behavior for small frequency. In fact, when the consistent mass matrix is used, they agree to order ω^2 (or $(kL)^4$) terms. As can be seen from Figure 5.10, however, beyond $kL \simeq 2$, there are significant differences because the conventional form behaves monotonically while the present form exhibits multiple zeros. For the conventional element to have these zeros it is necessary to piece many of them together as was demonstrated for the rod element.

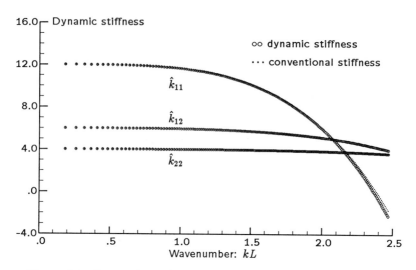

Figure 5.9: Comparison of conventional and spectral stiffnesses.

Throw-Off Element

The boundary conditions for the spectral beam element are quite similar to those of the conventional element. The new condition of interest is the throw-off bar. The stiffness relation is established by taking the shape function as

$$\hat{v}(x) = \mathbf{A}e^{-ikx} + \mathbf{B}e^{-kx}$$

which corresponds to a wave only moving outward and gives

$$\left\{ \begin{array}{c} \hat{V}_1 \\ \hat{m}_1 \end{array} \right\} = EI \left[\begin{array}{cc} (i-1)k^3 & ik^2 \\ ik^2 & (i+1)k \end{array} \right] \left\{ \begin{array}{c} \hat{v}_1 \\ \hat{\phi}_1 \end{array} \right\}$$

This, too, is complex and will cause the resulting structural stiffness matrix also to be complex.

Responses at Arbitrary Locations

The displacement formulation is such that the degrees of freedom are the basic unknowns and, for the first pass through, the reduced nodal quantities $\hat{v}$ and $\hat{\phi}$ are obtained. The unknown nodal loads for selected members are then calculated from

$$\{\hat{F}\} = [\ \hat{k}\ ]\{\hat{u}\}$$

The member loads and displacements are determined at arbitrary locations by appropriate differentiation of the shape functions taking the form

$$\hat{v}(x) = \mathbf{A}e^{-ikx} + \mathbf{B}e^{-kx} + \mathbf{C}e^{-ik(L-x)} + \mathbf{D}e^{-k(L-x)}$$

$$\mathbf{A} = \frac{1}{4}(\hat{v}_1 + i\frac{\hat{\phi}_1}{k} + \frac{\hat{m}_1}{EIk^2} - \frac{\hat{V}_1}{EIk^3}), \qquad \mathbf{B} = \frac{1}{4}(\hat{v}_1 - i\frac{\hat{\phi}_1}{k} - \frac{\hat{m}_1}{EIk^2} - \frac{\hat{V}_1}{EIk^3})$$

$$\mathbf{C} = \frac{1}{4}(\hat{v}_2 - i\frac{\hat{\phi}_2}{k} - \frac{\hat{m}_2}{EIk^2} - \frac{\hat{V}_2}{EIk^3}), \qquad \mathbf{D} = \frac{1}{4}(\hat{v}_2 + i\frac{\hat{\phi}_2}{k} + \frac{\hat{m}_2}{EIk^2} - \frac{\hat{V}_2}{EIk^3})$$

There is an alternative form to these that use the nodal displacements directly as in equation (5.21). It is computationally more intensive but it offers better numerical stability. The corresponding expressions for the responses in the throw-off bars are

$$\mathbf{A} = \frac{1}{(i-1)}\left[\hat{v}_1 + \frac{\hat{\phi}_1}{k}\right], \qquad \mathbf{B} = \frac{1}{(1+i)}\left[\hat{v}_1 - i\frac{\hat{\phi}_1}{k}\right], \qquad \mathbf{C} = \mathbf{D} = 0$$

It is possible to arrange for both left-handed and right-handed throw-off elements, but the simplest approach is to just arrange the nodal numbering so that all throw-off elements are right-handed.

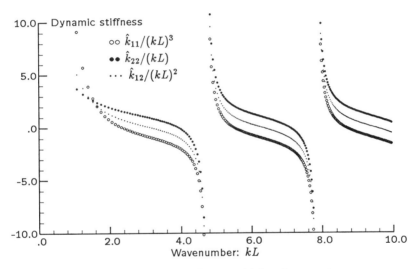

Figure 5.10: Spectral stiffness at higher frequencies.

5.6   Structural Formulation

The main motivation for using the matrix ideas of the finite element method is that once the basic building block of the element is established, then quite complicated problems can be solved simply by piecing together these simpler blocks. Precisely the same can be done with the spectral elements. Indeed, it is possible to convert a conventional finite element program to a spectral one without too much problem. This section points out some of the differences.

Assemblage

The dynamic structural stiffness matrix is assembled in a completely analogous way to that used for the static stiffness. The major difference is that it is done as part of the do-loop over all the frequency components. That is, the elements can be assembled into structural system form as

$$[\hat{K}]\{\hat{u}\} = \{\hat{P}\}$$

If the nodal displacement vector $\{\hat{u}\}$ is solved for at each frequency, then its time behavior is reconstructed simply by using the FFT inverse. Thus the problem can be thought of as a sequence of N pseudo-static problems. This adds an inefficiency in that the element stiffness must be evaluated and then assembled at each frequency. Further, the stiffness matrix can be complex and although it retains its symmetry and bandedness, this adds to the computer time for solution. As a compensation for this, the force history need not be specified (only its location) and therefore the same structural solution can be used for many different input histories.

A big advantage of the spectral formulation is that since there are a smaller number of nodes, then it is feasible to store the complete solution. For example, a plane frame with eleven members can have, say, seven nodes, giving a system size of twenty one. The storage requirements based on a 1024 point FFT transform and 4 byte words is about

$$21 \times 512 \times 4 \times 2 = 84\text{k bytes}$$

This very small size for the complete solution means it is now feasible to add a post-processing capability that allows the reading in of the nodal solution and obtaining the member loads or distributions at any desired point without having to redo the complete solution. This facility also allows solving inverse problems rapidly.

A prototype version of the above program coded for a personal computer gave typical run times on an 8 MHz AT type PC of about twenty minutes for a system of twenty degrees of freedom. This corresponds to the solution of a fairly complex frame comprised of ten members. The program was not fully optimized so it is believed this performance can be improved by a couple of factors. This program used for the studies reported in the next sections is not listed in the Appendix, but it is available through the author.

Discussion

The spectral approach to dynamics problems temporarily removes time from the description, thereby leaving a pseudo-static problem. Actually, there is one for each frequency component and often not much is gained thereby: solve the static problem in N time increments or solve the pseudo-static problem at N frequencies. This would be the case if the conventional system of equation (5.15) was treated by spectral analysis.

What makes the present approach so useful is that each element is solved exactly for its dynamics irrespective of its length. Thus the choice of element size is not restricted by the need to approximate the distributed mass (since that is done exactly) and allows the choice to be governed entirely by the structural connections and discontinuities. The added advantages of the spectral approach are that

- The effect of material damping and viscoelasticity can easily be incorporated simply by changing the spectrum relation.

- Higher-order beam or rod theories can be implemented without adding degrees of freedom to the system to be solved.

- Inverse problems can be performed. That is, if the response is known at some location then the disturbance causing it can be determined.

- The formulation is already in place for doing the usual type of spectral analysis and in particular, vibration analysis.

The disadvantage of the approach is that the dynamic stiffness exhibits a wide dynamic range thus putting extra demand on the computational numerics. Also, there is not the same programming support available as for the conventional formulation. It is expected, however, that both of these can be remedied in time.

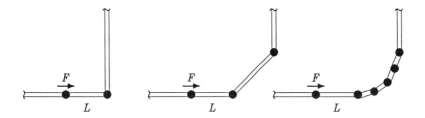

Figure 5.11: Various types of bends.

5.7   Structural Applications

The following examples are used to show the basic validity of the spectral formulation, and how it can complement the approaches of the other chapters. The first demonstrates the flexibility and versatility afforded by the matrix approach, as well as showing the inherent advantage of having a throw-off element. The second shows an application where resonance of the structure is important. The last example is the impact of a fairly complex structure.

Curved Rod

It was shown in Chapter 4 that the analysis of curved rods introduced coupled modes and complicated the analysis. Consequently, a great deal of time and effort must be expended in obtaining the spectrum relation. The present example shows an alternative approach — the curve is modeled as a sequence of straight elements, each of which has a simple description. The complexity of connecting them all together is handled by the matrix methodology.

To illustrate what happens, consider the sequence of three bends shown in Figure 5.11. They correspond to increasing smoothness in a 90° bend. The responses at the impact node are shown in Figure 5.12. The sharp bend shows that the incident pulse is reflected as if it met a free end, and very little of the energy is transmitted. When a 45° transition is introduced, distortion is observed in the reflected signal. This indicates mode conversion to flexural waves as well as more transmission of energy. It is also apparent that there are multiple reflections within the 45° member. The final case is a fairly smooth somewhat circular bend. This exhibits the smallest reflected initial pulse and a frequency analysis of it shows substantial filtering. The smoothness of the reflected portion of the signal indicates that the higher-frequency components are passing through unaffected.

It should be pointed out that while this matrix approach is useful for finite bends it would not do at all for, say, a helical spring. The analysis of Chapter 4 is then necessary since it would allow the study of very long bends. Also, it must be realized that there is no number of straight beam segments that will give the non-uniform stress and strain distribution of the curved beam.

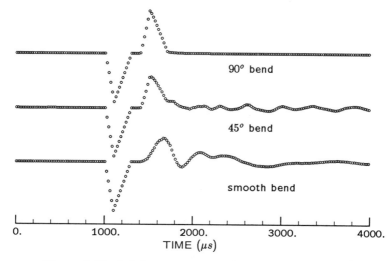

Figure 5.12: Axial response near the impact site.

Modal Analysis

While the spectral approach is good for dynamic wave propagation problems, it also has its use in conventional vibration analysis. This section shows a brief example. The portal frame shown in Figure 5.13 is the example structure because it is one often used to demonstrate the vibration of structures.

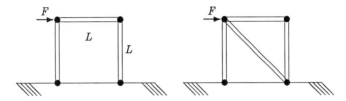

Figure 5.13: Portal frames.

The resonances can be determined by monitoring the behavior of the determinant of the system of equations. This is shown in Figure 5.14 with and without a cross-bar. Note that once the resonances have been located then precise values can be obtained by zooming on the frequencies. Doing this is only a matter of inputting a narrower range of frequencies with a finer increment. The ability to zoom is especially useful for the higher resonances because it does not require a finer mesh nor does it require the determination of the lower resonances. This is a decided advantage over the conventional finite element approach.

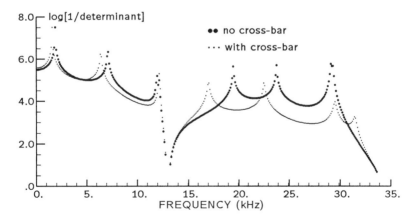

Figure 5.14: Resonances of the portal frame.

The first and second peaks shown in the figure are the first anti-symmetric and first symmetric modes, respectively. This is confirmed by comparison with the estimates given in Reference [233]. The formulas used are

$$\omega_1 = \frac{3.21}{L^2}\sqrt{\frac{EI}{\rho A}} = 1803 \text{ Hz}, \qquad \omega_2 = \frac{12.7}{L^2}\sqrt{\frac{EI}{\rho A}} = 6942 \text{ Hz}$$

The mode shapes are obtained by plotting the distribution between the nodal values. The nodal values themselves can be obtained by solving the system at the selected frequency.

It should be pointed out that there is a difference in doing the modal analysis and doing the forced response. For the modal analysis, the frequency range and spacing are entirely arbitrary. In fact, only one frequency need be analyzed if that is all that is desired. For the forced response, however, the frequencies are dictated by the requirements of the FFT inverse transform. For finite structures such as the portal frame, the propagating pulse has many reflections. Indeed, a true elastic system would continue this ad infinitum without diminish of amplitude. Unfortunately for the FFT formulation this would mean components propagating into neighboring time windows (as explained in Chapter 2). This can be alleviated somewhat in the present example by including material damping and having a very large time window. The net result of this is that a finer frequency resolution must be used and this is at the expense of a large N. The problem is less severe in more complex structures because the energy is naturally being diffused and will not be trapped so readily.

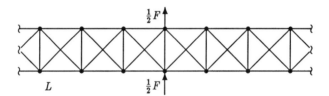

Figure 5.15: Truss structure.

Impact of a Truss Structure

This final example shows how a complicated truss structure can exhibit global behavior quite similar to one of the elementary structural models. The truss is shown in Figure 5.15, but the actual system solved was reduced by imposing symmetry about the impact line. This also shows how it is possible to impose more than one force at a time (although in this case they both have the same histories.)

Figure 5.16 shows the axial forces in the four members connected to the center bar. It is obvious that there is anti-symmetry in the loads with predominant tension

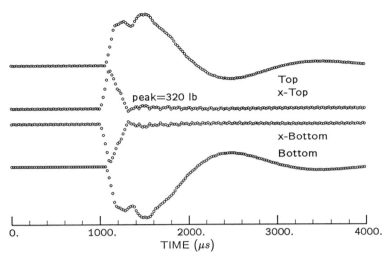

Figure 5.16: Axial forces in the members of the first cell.

on the top and predominant compression on the bottom. This is exactly how a beam would behave if struck in a similar manner.

Figure 5.17 shows how the axial force varies away from the impact. It exhibits the dispersive characteristics of a beam with the load at the center experiencing the maximum value and longest duration. The other members show more of the oscillations through zero, as well as diminishing amplitude that is also typical of beams.

There are many variations on these problems that can be pursued as well as phenomena that can be investigated. For example, it is apparent from Figures 5.16 and 5.17 that a good deal of energy is trapped in the horizontal members near the impact. So an interesting question is how to design the joints so that this trapping does not occur and the energy can diffuse throughout the structure. Such a design would tend to minimize local impact damage. On the other hand, some structural designs might desire to trap as much energy as possible so as to minimize damage to the remainder of the structure.

Global Energy Transport

An aspect of these problems not covered here (but should eventually be addressed) is that of describing the global flow of energy through the system. Keep in mind that while the spectral approach is capable of describing in detail the motion of every particle in the structure, the fundamental type of information recorded are the spectral components of the nodal displacements. It would be of interest, therefore, to investigate how this global information (which is already in a convenient reduced

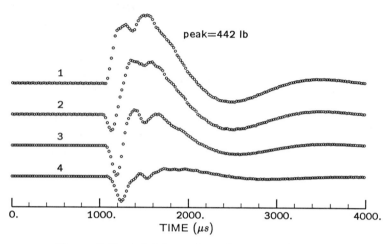

Figure 5.17: Axial forces in each top member.

form) could be made to give a big picture view of energy propagation. This integrated level is somewhat similar to the modal analysis view of structural dynamics except that, fundamentally, it is based on a transient event and not the long-term steady-state behavior. In other words, the inter-component phase relation is an important part of the picture.

Chapter 6

Waves in Two Dimensions

The analysis of frames, in a way, provided an extension of the one-dimensional analysis to multiple dimensions. By facilitating the description of multiple connected members, it allowed a global analysis. The waves in each case are confined to the waveguide and remain one dimensional — only the frame members were multidimensional. Of interest in this chapter is another form of multidimensionality where the medium itself can support a multidimensional wave. The two particular examples considered are that of in-plane waves and flexural waves in flat plates. These, respectively, are the two-dimensional equivalents of waves in rods and beams. They will also help in the understanding of the reduction of the general theory to the simpler one-dimensional waveguides.

This is a difficult and mathematically cumbersome area so only a limited number of aspects will be treated. Perhaps the most striking difference between the two-dimensional analysis and waveguides is the number of modes possible in each case. An attempt is made to show how the modes of both are related. Another different aspect of importance is that impacts usually result in non-planar waves. That is, for isotropic plates, say, the waves are circularly crested, and this gives the opportunity to introduce non-cartesian coordinates as well as the double series approach to problems. This is also used as the example to demonstrate the spectral analysis as a tool in force identification.

6.1   Waves in Infinite Media

The simplest exact solutions available are when the waves are propagating in an unbounded medium and these will be treated first. This introduces the P and S waves. To further aid in systematizing the results, the Helmholtz decomposition is also used.

For convenience, the indicial notation is sometimes used where $x_i = (x_1, x_2, x_3) = (x, y, z)$ and a subscript comma indicates partial differentiation. It may also be remarked that the exact equations of the theory of elasticity are used.

Navier's Equations and Helmholz Potentials

The development follows analogously to that used for rods and beams. First, the equations of motion are

$$\sum_j \frac{\partial \sigma_{ij}}{\partial x_j} + \rho f_i = \rho \ddot{u}_i$$

where f_i are the components of the body force. The relation between strain, rotation, and displacement is

$$2e_{ij} = \frac{\partial u_i}{\partial x_j} + \frac{\partial u_j}{\partial x_i}, \qquad 2\omega_{ij} = \frac{\partial u_i}{\partial x_j} - \frac{\partial u_j}{\partial x_i}$$

And finally, the elastic Hooke's law is taken in the form

$$\sigma_{ij} = 2\mu e_{ij} + \lambda \sum_k e_{kk}\delta_{ij}$$

where μ and λ are the Lamé constants and δ_{ij} is the Kronecker delta. Combining all these and using the displacement as the dependent variable gives the Navier's equations as

$$(\lambda + \mu) \sum_k u_{k,ki} + \mu \sum_k u_{i,kk} + \rho f_i = \rho \ddot{u}_i \tag{6.1}$$

This is a set of three coupled equations with second-order derivatives in both space and time.

The Helmholz decomposition represents the displacements by

$$u_i = \frac{\partial \Phi}{\partial x_i} + \sum_p \sum_q \epsilon_{ipq} \frac{\partial H_q}{\partial x_p}, \qquad \sum_k \frac{\partial H_k}{\partial x_k} = 0 \tag{6.2}$$

where ϵ_{ipq} is the permutation symbol. This represents the three displacements u_1, u_2, u_3 with the four potential functions Φ, H_1, H_2, H_3. Note that the extra condition uniquely determines H_q. On substituting for u_i into the Navier's equations, get on rearranging

$$\frac{\partial}{\partial x_i}[c_P^2 \nabla^2 \Phi - \ddot{\Phi}] + \sum_p \sum_q \frac{\partial}{\partial x_p}[c_S^2 \nabla^2 H_q - \ddot{H}_q]\epsilon_{pqi} = 0$$

where the Laplacian is defined as

$$\nabla^2 \equiv \frac{\partial^2}{\partial x_1^2} + \frac{\partial^2}{\partial x_2^2} + \frac{\partial^2}{\partial x_3^2} = \frac{\partial^2}{\partial x^2} + \frac{\partial^2}{\partial y^2} + \frac{\partial^2}{\partial z^2}$$

These equations are satisfied if the terms inside the square brackets are zero, that is, if Φ and H_q are chosen such that

$$c_P^2 \nabla^2 \Phi - \ddot{\Phi} = 0, \qquad c_S^2 \nabla^2 H_q - \ddot{H}_q = 0 \tag{6.3}$$

These are wave equations and it will be shown in more detail later that Φ is associated with the P-wave that travels with speed c_P, given by

$$c_P^2 = \frac{\lambda + 2\mu}{\rho} = \frac{E(1-\nu)}{\rho(1+\nu)(1-2\nu)} = c_o^2 \frac{1-\nu}{(1+\nu)(1-2\nu)} \tag{6.4}$$

These are called dilational, irrotational, longitudinal, voluminal, or primary (P) waves. Note that c_P is always greater than c_o with the special cases

$$
\begin{array}{llll}
\text{if} & \nu = 0\,, & \text{then} & c_P = c_o \\
\text{if} & \nu = .25\,, & \text{then} & c_P = 1.1 c_o
\end{array}
$$

The other three potentials, H_q, are associated with the S-wave that travels with speed c_S given by

$$c_S^2 = \frac{\mu}{\rho} = \frac{G}{\rho} = \frac{E}{2\rho(1+\nu)} = c_o^2 \frac{1}{2(1+\nu)} \tag{6.5}$$

These are called equi-voluminal, distortional, rotational, shear, or secondary (S) waves. Note that c_S is always less than c_o with the special cases

$$
\begin{array}{llll}
\text{if} & \nu = 0\,, & \text{then} & c_S = 0.71 c_o \\
\text{if} & \nu = .25\,, & \text{then} & c_S = 0.63 c_o
\end{array}
$$

It is quite obvious that both of these waves are non-dispersive.

It is worth pointing out that it is interactions with boundaries that usually makes the waves dispersive. In the construction of a waveguide theory, the boundary conditions are usually already embedded in the theory and that is why they are usually dispersive.

Waves in One Plane

Without loss of generality, consider only plane waves propagating in the $x-y$ plane, that is, assume there is no z dependence. Rather than write out the stresses in terms of Φ and H_q simultaneously, it is more convenient to separate them as follows.

Φ only

The differential equation to be satisfied is

$$c_P^2 \left\{ \frac{\partial^2}{\partial x^2} + \frac{\partial^2}{\partial y^2} \right\} \Phi - \ddot{\Phi} = 0 \tag{6.6}$$

The displacements are

$$u_x = \Phi_{,x}\,, \qquad u_y = \Phi_{,y}\,, \qquad u_z = 0 \tag{6.7}$$

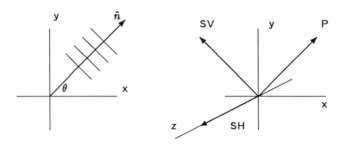

Figure 6.1: Directions of the P, SV, and SH waves.

This is a displacement vector in the $x - y$ plane pointing in a direction normal to the surface represented by the function $\Phi(x,y,t)$ as shown in Figure 6.1. It is also perpendicular to the P-wave. The stresses are

$$
\begin{aligned}
\sigma_{xx} &= (\lambda + 2\mu)\nabla^2\Phi - 2\mu\Phi_{,yy} \\
\sigma_{yy} &= (\lambda + 2\mu)\nabla^2\Phi - 2\mu\Phi_{,xx} \\
\sigma_{zz} &= \lambda\nabla^2\Phi \\
\sigma_{xy} &= 2\mu\Phi_{,xy} \\
\sigma_{xz} &= \sigma_{yz} = 0
\end{aligned}
\tag{6.8}
$$

This behavior is analogous to the one-dimensional longitudinal wave in a rod.

H_z only

The differential equation to be satisfied is

$$
c_S^2\left\{\frac{\partial^2}{\partial x^2} + \frac{\partial^2}{\partial y^2}\right\} H_z - \ddot{H}_z = 0
\tag{6.9}
$$

The displacements are

$$
u_x = H_{z,y}, \qquad u_y = -H_{z,x}, \qquad u_z = 0
\tag{6.10}
$$

This also is a displacement vector in the $x - y$ plane, but is pointing in a direction tangential to the surface represented by $H_z(x,y,t)$ as shown in Figure 6.1. It is therefore called an SV-wave. The stresses are

$$
\begin{aligned}
\sigma_{xx} &= 2\mu H_{z,xy} \\
\sigma_{yy} &= -2\mu H_{z,xy} \\
\sigma_{xy} &= \mu[-H_{z,xx} + H_{z,yy}] \\
\sigma_{zz} &= \sigma_{xz} = \sigma_{yz} = 0
\end{aligned}
\tag{6.11}
$$

The normal stresses indicate a state of pure shear.

H_x and H_y only

The differential equations to be satisfied are

$$c_S^2 \left\{ \frac{\partial^2}{\partial x^2} + \frac{\partial^2}{\partial y^2} \right\} H_x - \ddot{H}_x = 0$$

$$c_S^2 \left\{ \frac{\partial^2}{\partial x^2} + \frac{\partial^2}{\partial y^2} \right\} H_y - \ddot{H}_y = 0$$

$$\frac{\partial H_x}{\partial x} + \frac{\partial H_y}{\partial y} = 0 \tag{6.12}$$

These equations, of course, are coupled. The displacements are

$$u_x = u_y = 0, \qquad u_z = -H_{x,y} + H_{y,x} \tag{6.13}$$

This displacement vector is pointing only in the z direction, perpendicular to both the P and SV waves (as shown in Figure 6.1). That is, it is a shear in the horizontal direction and is therefore called an SH-wave.

The stresses are

$$\sigma_{xx} = \sigma_{yy} = \sigma_{zz} = \sigma_{xy} = 0$$

$$\sigma_{xz} = \mu[-H_{x,xy} + H_{y,xx}] = \mu \nabla^2 H_y$$

$$\sigma_{yz} = \mu[-H_{x,yy} + H_{y,xy}] = -\mu \nabla^2 H_x \tag{6.14}$$

indicating the presence of only shear stresses.

This decomposition of the waves helps in the bookkeeping when trying to solve boundary value problems. Some examples of this follow.

6.2 Semi-infinite Media

Only two waves, a P-wave and an S-wave, can propagate in the infinite medium. When a wave interacts with a boundary, however, new waves are generated. Some of them can be interpreted as just reflections of the incident wave, but (as will be shown), there are also totally new waves generated that are associated with the surface. In seismology, for example, it is observed that a third wave always follows the arrival of the P and S waves. Rayleigh postulated that perhaps there were surface waves generated. More details can be found in Reference [81].

Some study is needed of these surface waves in order to understand the effect of boundary surfaces on wave propagation in waveguides.

Surface Potentials

The first step is to construct a set of solutions appropriate for surfaces. Let the wave be propagating in the x-direction parallel to the surface, and of the form

$$\Phi(x, y, t) = f(y) e^{-i(kx - \omega t)}$$

Figure 6.2: Geometry for half-space and doubly bounded space.

The function $f(y)$ is determined by substituting into the wave equation (6.6) to get

$$\frac{\partial^2 f}{\partial y^2} - \left[k^2 - \left(\frac{\omega}{c_P}\right)^2\right]f = 0$$

This is an equation with constant coefficients and so has exponentials as a solution. Consequently, the potential function becomes

$$\Phi = \mathbf{A}e^{i\eta y}e^{-i(kx-\omega t)}\,,\qquad \eta^2 = (\omega/c_P)^2 - k^2 \equiv k_P^2 - k^2 \tag{6.15}$$

Note that if k is greater than k_P, then η is imaginary and the solution is a wave in x but evanescent in y. That is, it is localized to near $y = 0$. Recall that, as yet, k is undetermined. A similar analysis for the H_z potential gives

$$H_z = \mathbf{B}e^{i\bar\eta y}e^{-i(kx-\omega t)}\,,\qquad \bar\eta^2 = (\omega/c_S)^2 - k^2 \equiv k_S^2 - k^2 \tag{6.16}$$

Note that it is possible for both η and $\bar\eta$ to be complex as well as being negative.

The spectral form for the displacements corresponding to these potentials are

$$\begin{aligned}
\hat{u}_x &= -ik\mathbf{A}\hat\Phi + i\bar\eta\mathbf{B}\hat{H}_z\\
\hat{u}_y &= i\eta\mathbf{A}\hat\Phi + ik\mathbf{B}\hat{H}_z\\
\hat{u}_z &= 0
\end{aligned} \tag{6.17}$$

where $\hat\Phi = e^{i\eta y}e^{-ikx}$ and $\hat{H}_z = e^{i\bar\eta y}e^{-ikx}$. The stresses corresponding to these are

$$\begin{aligned}
\hat\sigma_{xx} &= [(\lambda+2\mu)(-k^2-\eta^2)+2\mu\eta^2]\mathbf{A}\hat\Phi + 2\mu[k\bar\eta]\mathbf{B}\hat{H}_z\\
\hat\sigma_{yy} &= [(\lambda+2\mu)(-k^2-\eta^2)+2\mu k^2]\mathbf{A}\hat\Phi - 2\mu[k\bar\eta]\mathbf{B}\hat{H}_z\\
\hat\sigma_{xy} &= 2\mu[k\eta]\mathbf{A}\hat\Phi + \mu[k^2-\bar\eta^2]\mathbf{B}\hat{H}_z
\end{aligned} \tag{6.18}$$

Similar expressions can be obtained for the other solution by replacing η and $\bar\eta$ everywhere by $-\eta$ and $-\bar\eta$, respectively.

When imposing the boundary conditions, it is required that the phases of Φ and H_z be the same. This is ensured by having $(kx - \omega t)$ common in both functions.

Mixed Boundary Condition

First consider the somewhat artificial boundary conditions at $y = 0$ of $u_y = 0$, $\sigma_{xy} = 0$ giving

$$\begin{bmatrix} i\eta & ik \\ 2k\eta & (k^2 - \bar{\eta}^2) \end{bmatrix} \left\{ \begin{matrix} A \\ B \end{matrix} \right\} = 0 \tag{6.19}$$

For a non-trivial solution for the amplitude coefficients A, B, the determinant must be zero, giving

$$i\eta(-k^2 - \bar{\eta}^2) = 0 \tag{6.20}$$

Since both η and $\bar{\eta}$ are related to the wavenumber k, then this allows k to be determined. The only situation when this equation is satisfied is when

$$\eta = 0 \qquad \Rightarrow \qquad k = k_P = \omega/c_P \tag{6.21}$$

This shows, on taking $B = (\eta/k)A = 0$, that the second wave is not required in order to satisfy the boundary conditions.

Traction Free Boundary — Rayleigh Surface Waves

As another case, consider when the surface tractions are zero. That is, on the surface $y = 0$, impose that both tractions be zero for any x. This traction-free condition now becomes

$$\begin{bmatrix} [-(\lambda + 2\mu)(k^2 + \eta^2) + 2\mu k^2] & 2\mu[-k\bar{\eta}] \\ 2\mu[k\eta] & \mu[k^2 - \bar{\eta}^2] \end{bmatrix} \left\{ \begin{matrix} A \\ B \end{matrix} \right\} = 0 \tag{6.22}$$

This can be simplified somewhat to

$$\begin{bmatrix} (2k^2 - k_S^2) & -2k\bar{\eta} \\ 2k\eta & (2k^2 - k_S^2) \end{bmatrix} \left\{ \begin{matrix} A \\ B \end{matrix} \right\} = 0 \tag{6.23}$$

Again, for a nontrivial solution for the amplitude coefficients the determinant must be zero. That is

$$(2k^2 - k_S^2)^2 + 4k^2\eta\bar{\eta} = 0$$

Substituting for η and $\bar{\eta}$ in terms of k and rearranging gives

$$16(-1 + \gamma^2)\left(\frac{k}{k_S}\right)^6 + 8(3 - 2\gamma^2)\left(\frac{k}{k_S}\right)^4 - 8\left(\frac{k}{k_S}\right)^2 + 1 = 0 \tag{6.24}$$

where γ is taken as

$$\gamma^2 = \frac{\mu}{\lambda + 2\mu} = \frac{c_S^2}{c_P^2} = \frac{(1 - 2\nu)}{2(1 - \nu)}$$

There is no explicit frequency occurring in the above characteristic equation (because there are only ratios of wavenumbers); consequently, the surface wave is non-dispersive. Also, it is cubic in k^2, hence there are three possible modes.

For the special case of $\lambda = \mu$, ($\nu = 0.25$) then $\gamma^2 = 1/3$ and the equation can be factored as

$$\left[4\left(\frac{k}{k_S}\right)^2 - 1\right]\left[4\left(\frac{k}{k_S}\right)^2 - 3 + \sqrt{3}\right]\left[4\left(\frac{k}{k_S}\right)^2 - 3 - \sqrt{3}\right] = 0$$

This gives three possibilities for the wavenumber and speed. A particular case of interest is when the wave is localized near the surface but still propagating. This occurs when both η and $\bar{\eta}$ are imaginary. That is, when

$$k^2 > k_P^2 = \frac{\omega^2}{c_P^2} = k_S^2 \gamma^2 \qquad \text{and} \qquad k^2 > k_S^2 = \frac{\omega^2}{c_S^2}$$

The second condition dominates, hence it is necessary to choose

$$\frac{k}{k_S} = \tfrac{1}{2}\sqrt{3 - \sqrt{3}} = 1.08 \tag{6.25}$$

This special wave is called a *Rayleigh wave* and the speed is denoted by c_R. The potentials and amplitude ratios are

$$\begin{aligned}
\Phi &= A e^{\sqrt{1-(c_R/c_P)^2}\,ky}\, e^{-ik(x-c_Rt)} \\
H_z &= B e^{\sqrt{1-(c_R/c_P)^2}\,ky}\, e^{-ik(x-c_Rt)} \\
\frac{B}{A} &= \frac{2i\sqrt{1-(c_R/c_P)^2}}{2-(c_R/c_S)^2}
\end{aligned} \tag{6.26}$$

This is a non-dispersively propagating wave whose depth of penetration in the y-direction depends on the wavenumber k and, consequently, on the frequency ω also. More aspects of this wave are discussed fully in Reference [223]. The above is for a specific value of Poisson's ratio, in general, the wave speed is related to ν ratio by

$$\frac{k_S}{k} = \frac{c}{c_S} \approx \frac{0.87 + 1.12\nu}{1+\nu} \tag{6.27}$$

Rayleigh waves are usually caused by some impact or due to non-planar waves incident on a surface. Lamb in Reference [133] considered this quite extensively in 1904. A more recent treatment can be found in References [196,198]. The significance of these results is to show how a boundary can cause totally new waves to be generated, and that these waves are unlike the incident waves. This, of course, was also observed in the last three chapters, but here the effect on the basic wave itself is being demonstrated. It may help to phrase this in terms of the previous chapters. That is, the H_z set of solutions must be introduced to help satisfy the boundary conditions. As a result, the wavenumber has three families of values (or three modes). One of these modes corresponds to a wave propagating along the surface and decaying exponentially into the medium. Thus the presence of this wave is really an artifact of coupling.

6.3   Doubly Bounded Media

To illustrate the problems associated with obtaining exact solutions when there are multiple boundaries, the case of symmetric Lamb waves in a plate will be treated. These are similar to longitudinal waves traveling in the plane of a plate. It will also give an idea of how the approximate theories relate to the exact solutions. Of interest is the case where the top and bottom surfaces are traction free as shown in Figure 6.2.

Symmetric Potentials

Again, the first step is to construct the potentials appropriate for the medium. Assume plane wave motion in the x-direction and let

$$\Phi(x,y,z,t) = Y_P(y)e^{-i(kz-\omega t)}$$
$$H_z(x,y,z,t) = Y_S(y)e^{-i(kz-\omega t)}$$

By substitution into the respective wave equations (6.6 & 6.9) specific forms for Y_P and Y_S are obtained as

$$Y_P(y) = \mathbf{A}\cos(\eta y), \qquad Y_S(y) = \mathbf{B}\sin(\bar{\eta}y) \qquad (6.28)$$

provided η and $\bar{\eta}$ are as given before. The displacements corresponding to these are

$$\begin{aligned}
u_x &= [-ik\mathbf{A}\cos(\eta y) + \bar{\eta}\mathbf{B}\cos(\bar{\eta}y)]e^{-i(kz-\omega t)}\\
u_y &= [-\eta\mathbf{A}\sin(\eta y) + ik\mathbf{B}\sin(\bar{\eta}y)]e^{-i(kz-\omega t)}\\
u_z &= 0
\end{aligned} \qquad (6.29)$$

Thus, it is apparent that the specific choice of Y_P and Y_S gives displacements in the x-direction that are symmetric with respect to y. The stresses corresponding to these displacements are

$$\begin{aligned}
\sigma_{zz} &= [-(2\mu k^2 + k_p^2)\mathbf{A}\cos\eta y - 2\mu(ik\bar{\eta})\mathbf{B}\cos\bar{\eta}y]e^{-i(kz-\omega t)}\\
\sigma_{yy} &= [\mu(k^2-\eta^2)\mathbf{A}\cos(\eta y) + 2\mu(ik\bar{\eta})\mathbf{B}\cos(\bar{\eta}y)]e^{-i(kz-\omega t)}\\
\sigma_{zy} &= [\mu(2ik\eta)\mathbf{A}\sin(\eta y) - \mu(\bar{\eta}^2-k^2)\mathbf{B}\sin(\bar{\eta}y)]e^{-i(kz-\omega t)}
\end{aligned} \qquad (6.30)$$

and it is noted that the longitudinal stress σ_{zz} is symmetric in y also.

Mixed Boundary Condition

As in the last section, first consider the somewhat artificial boundary conditions at $y = \pm h/2$ of $u_y = 0$, $\sigma_{zy} = 0$ giving

$$\begin{bmatrix} -\eta\sin(\eta h/2) & ik\sin(\bar{\eta}h/2) \\ 2ik\eta\sin(\eta h/2) & -(\bar{\eta}^2 - k^2)\sin(\bar{\eta}h/2) \end{bmatrix}\begin{Bmatrix} \mathbf{A} \\ \mathbf{B} \end{Bmatrix} = 0 \qquad (6.31)$$

The determinant must be zero giving, after some simplification,

$$\eta(k_S^2)\sin(\eta h/2)\sin(\bar{\eta}h/2) = 0 \tag{6.32}$$

There are two situations where this equation is satisfied

$$\eta = \frac{m\pi}{h} \quad \Rightarrow \quad k = \sqrt{\left(\frac{\omega}{c_P}\right)^2 - \left(\frac{m\pi}{h}\right)^2}, \qquad m = 0, 2, 4, ...$$

$$\bar{\eta} = \frac{n\pi}{h} \quad \Rightarrow \quad k = \sqrt{\left(\frac{\omega}{c_S}\right)^2 - \left(\frac{n\pi}{h}\right)^2}, \qquad n = 0, 2, 4, ... \tag{6.33}$$

A plot of these spectrum relations are shown in Figure 6.3. The cut-off frequencies for the two families are obviously

$$\omega_0 = \frac{m\pi}{h}c_P, \qquad\qquad m = 0, 2, 4....$$

$$\omega_0 = \frac{n\pi}{h}c_S, \qquad\qquad n = 2, 4.... \tag{6.34}$$

The lowest mode $(m = 0)$ corresponds to longitudinal waves. When the plate thickness h is very small, then the characteristic equation becomes $\eta^2\bar{\eta} = 0$. This appears to give two spectrum relations, but $\bar{\eta}$ is trivial, hence there is only one mode and is simply

$$k = k_p$$

This is interpreted as a wave traveling at the longitudinal wave speed.

The amplitude ratio at each mode is

$$\mathbf{B} = \frac{\eta\sin(\eta h/2)}{ik\sin(\bar{\eta}h/2)}\mathbf{A} \tag{6.35}$$

This shows that for each mode the P and SV behavior is uncoupled. For example, for the $m = 0, 2, 4....$ family

$$u_x = -\mathbf{A}ik\cos\left(\frac{m\pi}{h}y\right)e^{-i(k_m x - \omega t)}, \qquad u_y = -\mathbf{A}\frac{m\pi}{h}\sin\left(\frac{m\pi}{h}y\right)e^{-i(k_m x - \omega t)}$$

This example shows that the solution has an infinite number of modes. Also the shear and longitudinal modes are uncoupled.

Traction Free Boundaries – Lamb Waves

The more realistic boundary condition of a free surface at $y = \pm h/2$ is treated by imposing the requirements that $\sigma_{yy} = 0$ and $\sigma_{xy} = 0$ for all x. This gives

$$\begin{bmatrix} (k^2 - \eta^2)\cos(\eta h/2) & 2ik\bar{\eta}\cos(\bar{\eta}h/2) \\ 2ik\eta\sin(\eta h/2) & (k^2 - \bar{\eta}^2)\sin(\bar{\eta}h/2) \end{bmatrix}\begin{Bmatrix} \mathbf{A} \\ \mathbf{B} \end{Bmatrix} = 0 \tag{6.36}$$

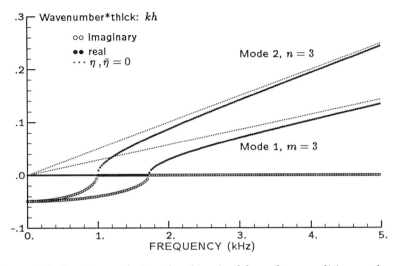

Figure 6.3: Spectrum relations for the mixed boundary condition modes.

The determinant of this must be zero in order to have a nontrivial solution. This gives the characteristic equation which, on rearranging, is

$$\tan(\eta h/2) = -\frac{(k^2 - \eta^2)(k^2 - \bar{\eta}^2)}{4k^2 \eta \bar{\eta}} \tan(\bar{\eta} h/2)$$

Substituting the expressions for η and $\bar{\eta}$ gives

$$\tan(\sqrt{k_P^2 - k^2}\, h/2) = \frac{-(2k^2 - k_P^2)(2k^2 - K_S^2)}{4k^2 \sqrt{(k_P^2 - k^2)(k_S^2 - k^2)}} \tan(\sqrt{k_S^2 - k^2}\, h/2)$$

This is a complicated transcendental equation. The relation between wavenumber k and frequency ω (through k_P and k_S) is obviously multivalued and must be obtained numerically. The amplitude ratio at each mode is

$$\frac{\mathbf{B}}{\mathbf{A}} = \frac{-(k^2 - \bar{\eta}^2)\cos(\eta h/2)}{2ik\eta \cos(\bar{\eta} h/2)} \qquad (6.37)$$

This shows that for each mode the P and S behavior is coupled and probably no longer have their distinctive characteristics.

When the thickness h is very small, the above simplifies to

$$k^2 = \frac{k_P^2 k_S^2}{2(k_S^2 - k_P^2)} \qquad \text{or} \qquad k = \frac{\omega}{c_o}\sqrt{(1 + \nu)(1 - 2\nu)}$$

This is the non-dispersive relation for a longitudinal wave in a rod modified for plane strain.

General Solutions

To solve a problem with finite boundaries requires first the evaluation of all the spectrum relations, and then to sum up all the modes so as to be able to satisfy the boundary conditions. Examples of doing this can be found in References [51,94,160]. An exact theory with some fairly complete solutions is that of waves in circular bounded media (rods) and is known as the Pochhammer-Chree theory. The solutions have Bessel functions in the radial directions and very complicated spectrums. More details can be found in References [55,120,177,238]. To date, it seems the most useful contribution of these exact theories is in the evaluation of the approximate approaches. However, with the wider availability of computers they may have a resurgence, although it is beyond the capability of present day personal computers.

The following sections consider a variation on the above where the spectrum relations are relatively simple allowing concentration on the aspects of solving the boundary value problem itself. Some of the advantages and drawbacks of using spectral analysis for these problems are shown.

6.4 Flexural Behavior of Plates

Plates are the two-dimensional equivalent of beams and classical plate theory is its equivalent of the Bernoulli-Euler beam theory. Needless to say the behaviors of plates and beams are very similar, so only those aspects that are significantly different will be covered in this chapter.

The theory derived here is the 2-D equivalent of the Bernoulli-Euler beam theory. The only real difference is that resultants act on four sides of the plate element instead of two of the beam. Consequently, the derivation of the equations of motion will follow a similar approach as that used in the chapter on beams and so some of the steps will be abbreviated.

Equation of Motion

Let Q_x and Q_y be the resultant shear forces per unit length, and M_x, M_y, and M_{xy} be the resultant moments per length as shown in Figure 6.4. Neglecting rotational inertia effects, the equations of motion in the z-direction and about the x and y axes are, respectively,

$$\frac{\partial Q_x}{\partial x} + \frac{\partial Q_y}{\partial y} = \rho h \frac{\partial^2 w}{\partial t^2} - P$$

$$\frac{\partial M_x}{\partial x} + \frac{\partial M_{xy}}{\partial y} + Q_x = 0$$

$$\frac{\partial M_y}{\partial y} + \frac{\partial M_{yx}}{\partial x} + Q_y = 0 \qquad (6.38)$$

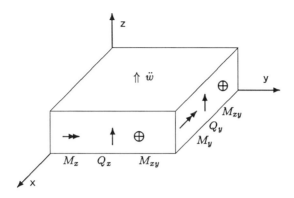

Figure 6.4: Element of stressed plate.

where h is the plate thickness and P is the applied load. The moments are related to the deflections by

$$M_x = D\left[\frac{\partial^2 w}{\partial x^2} + \nu\frac{\partial^2 w}{\partial y^2}\right]$$

$$M_y = D\left[\frac{\partial^2 w}{\partial y^2} + \nu\frac{\partial^2 w}{\partial x^2}\right]$$

$$M_{xy} = M_{yx} = D(1-\nu)\frac{\partial^2 w}{\partial x\partial y} \tag{6.39}$$

where $D = Eh^3/(1-\nu^2)$ is the stiffness and ν is Poisson's ratio. These allow the equations of motion to be combined and rewritten as

$$D\nabla^2\nabla^2 w + \rho h\frac{\partial^2 w}{\partial t^2} = P, \qquad \nabla^2 \equiv \frac{\partial^2}{\partial x^2} + \frac{\partial^2}{\partial y^2} \tag{6.40}$$

There is, of course, no D'Alembert solution to the homogeneous part of this equation. The functional representation for ∇^2 is used because the same identical equation is also valid in polar coordinates.

Spectral Analysis

Let the displacement history at an arbitrary point in the plate have the spectral representation

$$w(x,y,t) = \sum_n \hat{w}(x,y,\omega_n)e^{i\omega_n t} \tag{6.41}$$

The differential equation governing each Fourier coefficient $\hat{w}_n$, therefore, is

$$\{D\nabla^2\nabla^2 - \rho h\omega_n^2\}\hat{w}_n = 0 \tag{6.42}$$

In the remainder of the development the subscript n will be assumed and hence not written out explicitly. The solution of this equation can be written as linear sums of solutions of the following two differential equations

$$\nabla^2 \hat{w}_1 + \beta^2 \hat{w}_1 = 0\,, \qquad \nabla^2 \hat{w}_2 - \beta^2 \hat{w}_2 = 0\,, \qquad \beta^2 = \omega\sqrt{\frac{\rho h}{D}} \qquad (6.43)$$

These form the basic equations for further analysis and emphasize that there are two fundamentally different modes just as there is for beams.

There are many solutions which may be quite important when considering certain boundary effects. For example, consider the wave of the form

$$\hat{w}(x, y) = \tilde{w}(x)e^{i\eta y}$$

where assume that η is specified. The differential equation for $\tilde{w}$ becomes

$$\frac{d^2 \tilde{w}}{dx^2} + [-\eta^2 \pm \beta^2]\tilde{w} = 0$$

This has constant coefficients, hence, e^{-ikz} is a kernel solution. The spectrum relations are then

$$k = \pm\sqrt{\pm\beta^2 - \eta^2}$$

Specifically, suppose the wave is propagating in y and that $\eta = \beta \sin\theta = \beta S_\theta$ then

$$k = \pm\beta C_\theta\,, \qquad \pm\beta\sqrt{1 + S_\theta^2}$$

Thus, the possible waves are

$$\hat{w} = \mathbf{A}e^{-i\beta(\pm xC_\theta + yS_\theta)} + \mathbf{B}e^{\beta(\pm x\sqrt{1+S_\theta^2}+iyS_\theta)}$$

It is seen that the second attenuates in x but still propagates in y.

More generally, if η is taken to be real, then it is possible for k to transition from real to complex. This form will be used later in the section on double series solutions.

The relation between the wavenumber and frequency is given by

$$k = \pm\sqrt{\omega}\left[\frac{\rho h}{D}\right]^{\frac{1}{4}}, \qquad \pm i\sqrt{\omega}\left[\frac{\rho h}{D}\right]^{\frac{1}{4}} \qquad (6.44)$$

This is quite similar to that of the beam if the following associations are made

$$\frac{EI}{A} \Leftrightarrow \frac{D}{h} \quad \text{or} \quad E \Leftrightarrow \frac{E}{(1 - \nu^2)}$$

Thus the plate behaves as a beam in plane strain. This includes the expressions for the phase and group speeds also

$$c = \frac{\omega}{k} = \sqrt{\omega}\left[\frac{D}{\rho h}\right]^{\frac{1}{4}}, \qquad c_g = \frac{d\omega}{dk} = 2\sqrt{\omega}\left[\frac{D}{\rho h}\right]^{\frac{1}{4}} = 2c \qquad (6.45)$$

The behavior of Figure 4.3 for beams can be taken as the behavior of plates (including the higher-order effects shown in Figure 4.19).

6.5   Reflections from Boundaries

If the wave is incident normal to the boundary, then the plate generates a reflected wave and an attenuated wave localized at the boundary. However, when the wave is incident obliquely this wave can propagate along the boundary. This is the situation of interest here.

Boundary Conditions

Assume that the boundary is located at $x = 0$ and that the wave is incident from the third quadrant as shown in Figure 6.5. The type of boundary conditions to be satisfied are (according to Reference [32]) to be chosen from

$$\text{Displacement} \quad : \quad w = w(x, y, t)$$

$$\text{Slope} \quad : \quad \phi_x = \frac{\partial w}{\partial x}$$

$$\text{Moment} \quad : \quad M_x = D\left[\frac{\partial^2 w}{\partial x^2} + \nu\frac{\partial^2 w}{\partial y^2}\right]$$

$$\text{Shear} \quad : \quad V_x = -D\left[\frac{\partial^3 w}{\partial x^3} + (2 - \nu)\frac{\partial^3 w}{\partial x\partial y^2}\right]$$

Note that the Poisson's ratio ν enters the last two and acts to couple the variations in x to those in y. The shear to be specified is the Kirchhoff shear. This is the resultant shear and it is given by

$$\text{Kirchhoff Shear:} \quad V_x = Q_x - \frac{\partial M_{xy}}{\partial y} \tag{6.46}$$

Let the total solution be

$$w(x, y, t) = \sum\{\mathbf{A}e^{-i\beta x C_\theta} + \mathbf{B}e^{-\beta x C_\theta} + \mathbf{C}e^{i\beta x C_\theta} + \mathbf{D}e^{\beta z\sqrt{1+S_\theta^2}}\}e^{-i\beta y S_\theta}e^{i\omega t} \tag{6.47}$$

This gives a wave generally moving in the y-direction, with the $\mathbf{A}$ term being incident, the $\mathbf{C}$ term being reflected, and the $\mathbf{D}$ term being an x attenuated vibration. Since they have the common y term then the boundary conditions will be satisfied for any y.

A couple of boundaries are treated here so as to illustrate the approach. For simplicity, the incident $\mathbf{B}$ wave is assumed negligible.

Clamped Boundary

The boundary conditions at $x = 0$ give

$$\hat{w} = 0 = \mathbf{A} + \mathbf{C} + \mathbf{D}$$
$$\frac{\partial \hat{w}}{\partial x} = 0 = -i\beta C_\theta\mathbf{A} + i\beta C_\theta\mathbf{C} + \beta\sqrt{1 + S_\theta^2}\mathbf{D}$$

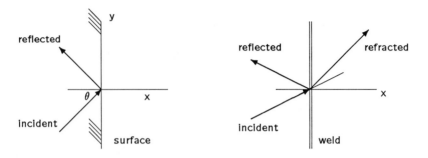

Figure 6.5: Wave reflections and transmissions near a straight edge.

giving the coefficients of the generated waves as

$$\mathbf{C} = -\frac{\sqrt{1 + S_\theta^2} + iC_\theta}{\sqrt{1 + S_\theta^2} - iC_\theta}\mathbf{A}, \qquad \mathbf{D} = \frac{2iC_\theta}{\sqrt{1 + S_\theta^2} - iC_\theta}\mathbf{A} \qquad (6.48)$$

If the angle of incidence, θ, is zero, then the reflected waves are phase shifted relative to the incident waves. On the other hand, if θ approaches $\pi/2$, then the $\mathbf{D}$ term disappears and $\mathbf{C} = -\mathbf{A}$ showing that the superposition is zero.

Elastic Boundary

Let the edge of the plate be supported by springs such that the boundary conditions become

$$M_x = -\alpha\frac{\partial w}{\partial x}, \qquad V_x = -Kw$$

where K and α are the linear and torsional spring constants respectively. In terms of the wave coefficients these become

$$D\beta^2\left\{-(C_\theta^2 + \nu S_\theta^2)\mathbf{A} + (1 + S_\theta^2 - \nu S_\theta^2)\mathbf{D} - (C_\theta^2 + \nu S_\theta^2)\mathbf{C}\right\}$$
$$= -\alpha\beta\left\{-iC_\theta\mathbf{A} + iC_\theta\mathbf{C} + \sqrt{l + S_\theta^2}\mathbf{D}\right\}$$

$$D\beta^3\left\{-iC_\theta\mathbf{A} + iC_\theta\mathbf{C} + \sqrt{l + S_\theta^2}\mathbf{D}\right\}\left\{1 - (2 - \nu)S_\theta^2\right\}$$
$$= -K\{\mathbf{A} + \mathbf{C} + \mathbf{D}\}$$

These are rearranged to give the amplitudes as

$$\begin{bmatrix} -1 + \nu^* + iC_\theta\alpha^* & 1 + \nu^* + \sqrt{1 + S_\theta^2}\,\alpha^* \\ iC_\theta(1 + \nu^*) - K^* & \sqrt{1 + S_\theta^2}(1 - \nu^*) + K^* \end{bmatrix}\left\{\begin{matrix} \mathbf{C} \\ \mathbf{D} \end{matrix}\right\} = \left\{\begin{matrix} 1 - \nu^* + iC_\theta\alpha^* \\ iC_\theta(1 + \nu^*) + K^* \end{matrix}\right\}\mathbf{A}$$
$$(6.49)$$

where

$$\nu^* = (1-\nu)S_\theta^2, \quad \alpha^* = \alpha/\beta D, \quad K^* = K/\beta^3 D$$

Obviously, the special cases of $K = 0$ or ∞ and $\alpha = 0$ or ∞ can easily be taken off. It is apparent that the general case is frequency dependent. It is interesting to note that the Poisson's ratio appears only in association with $\sin\theta$. This means that for normal incidence the Poisson's ratio does not play a role.

Reflection and Transmission at Stepped Plates

Plates often have stiffeners attached to them and it is of interest to know how they affect the wave. Again, the situation of most interest is when the wave is incident at an angle. Assume only an [A] wave is incident.

Let the waves in the two plates be

$$
\begin{aligned}
w_1(x,y,t) &= \sum \{ A_1 e^{-i\beta x C_{\theta 1}} + C_1 e^{i\beta x C_{\theta 1}} + D_1 e^{\beta x C_{\theta 1}} \} e^{-i\beta_1 y \sin\theta_1} e^{i\omega t} \\
w_2(x,y,t) &= \sum \{ A_2 e^{-i\beta x C_{\theta 2}} + B_2 e^{-\beta x C_{\theta 2}} \} e^{-i\beta_2 y \sin\theta_2} e^{i\omega t}
\end{aligned}
\tag{6.50}
$$

Assume the two plates are butt welded along the line $x = 0$ as shown in Figure 6.5. Let the weld be of zero mass and extent, then the continuity of displacement and slope at the joint requires for instance

$$
\begin{aligned}
w_1 &= w_2 = w_J \\
\frac{\partial w_1}{\partial x} &= \frac{\partial w_2}{\partial x} = \phi_J
\end{aligned}
$$

with similar ones for the moment and shear. Of interest here is the fact that in order for all these to be satisfied for all y, then the y phase terms must be equal. That is, it is required that

$$
\beta_1 \sin\theta_1 = \beta_2 \sin\theta_2 \quad \text{or} \quad \left[\frac{\rho h}{D}\right]_1^{\frac{1}{4}} \sin\theta_1 = \left[\frac{\rho h}{D}\right]_2^{\frac{1}{4}} \sin\theta_2
\tag{6.51}
$$

This is none other than Snell's law that is familiar in optics (see Reference [79]). Thus the incident wave is refracted in the second plate.

The simultaneous equations required to determine the amplitudes are

$$
\begin{bmatrix}
1 & 1 & -1 & -1 \\
iC_\theta & \sqrt{l+S_\theta^2} & ir_1 C_\theta & r_1\sqrt{l+S_\theta^2} \\
-1(1-\nu_1^*) & (1+\nu_1^*) & r_2(1-\nu_2^*) & -r_2(1+\nu_2^*) \\
-iC_\theta(1+\nu_1^*) & \sqrt{1+S_\theta^2}(1-\nu_1^*) & -ir_3 C_\theta(1+\nu_2^*) & r_3(1-\nu_2^*)
\end{bmatrix}
\begin{Bmatrix}
C_1 \\ D_1 \\ A_2 \\ B_2
\end{Bmatrix}
$$

$$
= \begin{Bmatrix}
-1 \\ iC_\theta \\ 1-\nu_1^* \\ iC_\theta(1+\nu_1^*)
\end{Bmatrix} A_1
\tag{6.52}
$$

where
$$r_1 = \beta^2/\beta_1, \qquad r_2 = (D_2\beta_2^2)/(D_1\beta_1^2), \qquad r_3 = r_1 r_2$$

Apart from the refraction, it is seen that the behavior is similar to that of equation (4.20) for beams.

The treatment of the stringer is approached the same as that of the elastic joint in beams. The waves on either side of it are as assumed before, but within the stringer itself the waves are taken as

$$w = \sum \{ \mathbf{A} e^{-i\beta x C_\theta} + \mathbf{B} e^{-\beta x C_\theta} + \mathbf{C} e^{i\beta x C_\theta} + \mathbf{D} e^{\beta x C_\theta} \} e^{-i\beta y S_\theta} e^{i\omega t}$$

That is, a standing wave is set up across the stringer while a propagating wave develops along its length. The relationship between the incident wave and the generated waves is now obtained by imposing continuity of displacement, slope, moment, and shear as was done for the beam. Of course, if the stringer is unbalanced with respect to the center line of the plate then longitudinal waves must also be added to the solution. The difficult part of the analysis is to obtain realistic estimates of the stringer parameter. That is, the model needs such quantities as stiffness, density and length, and it is necessary to convert the actual stringer into effective quantities. This is not done here but can be found in References [173].

6.6   Point Impact of a Plate

Let it be assumed that the response of an impacted plate can be represented adequately by classical plate theory. Since the impact is at a point, then the resultant wave fronts are circular and the field quantities are axisymmetric. This case serves as an example in the use of non-cartesian coordinates. It also demonstrates the use of spectral analysis in force determination or identification. An equivalent development for applied concentrated moments is shown in References [77,141].

Axisymmetric Equations

The differential operator for the spatial variation becomes under axisymmetric conditions
$$\nabla^2 = \frac{d^2}{dr^2} + \frac{1}{r}\frac{d}{dr}$$

The resulting equations are similar to that for plates on an elastic foundation and Reference [218] has shown that those problems have axisymmetric solutions of the form
$$\hat{w}(r) = \mathbf{A}J_0 + \mathbf{B}Y_0 + \mathbf{C}K_0 + \mathbf{D}I_0 \tag{6.53}$$

where J_0, Y_0, K_0, I_0 are Bessel functions with argument
$$z = \beta r, \qquad \beta = \sqrt{\omega}\left[\frac{\rho h}{D}\right]^{1/4}$$

and r is the radial distance from the plate center. The notation used for the Bessel functions is that of Reference [5] and a summary of their properties can be found in Appendix A. Let the load be applied at the center of the plate and imposing the following conditions:

- displacement is finite at $r = 0$

- displacement tends to zero at $r = \infty$

- displacement behaves as an outgoing wave at $r = \infty$

gives the kernel solution as

$$\hat{w} = \mathbf{A}\{J_0 - iY_0 - i\frac{2}{\pi}K_0\} \tag{6.54}$$

The remaining coefficient $\mathbf{A}$ is determined from the applied load condition. Let the force be a single harmonic and again following the approach in Reference [218] (for a static load) consider the equilibrium of a small cylinder section of the plate surrounding its point of application. The shear force per unit length Q, at the cylinder, is related to the force by

$$Q = \frac{F}{2\pi r} \quad \text{or} \quad \hat{Q} = \frac{\hat{F}}{2\pi r}$$

But the shear is also related to the displacement by

$$\hat{Q} = \beta^2 D \left[\hat{w}''' + \frac{1}{z}\hat{w}'' - \frac{1}{z^2}\hat{w}' \right] = \mathbf{A}\left\{ J_1 - iY_1 + \frac{2i}{\pi}K_1 \right\}$$

In the above the superscript prime indicates differentiation with respect to z. Letting r approach zero and equating both relations for $\hat{Q}$ allows $\mathbf{A}$ to be determined in terms of $\hat{F}$. The solution for the displacement now becomes

$$w(r,t) = \sum \frac{-i\hat{F}}{8\beta^2 D}\{J_0 - iY_0 - i\frac{2}{\pi}K_0\}e^{i\omega t} \tag{6.55}$$

The question of singular solutions is developed in good detail in Reference [115].

An interesting relation is obtained from the above by looking at the velocity at the impact site, i.e.,

$$\dot{w}(r = 0, t) = \sum \frac{\omega \hat{F}}{8\beta^2 D}\{J_0 - iY_0 - i\frac{2}{\pi}K_0\}e^{i\omega t} = \sum \frac{\hat{F}}{8\sqrt{\rho h D}}\{1\}e^{i\omega t}$$

Since the frequency does not occur in the transfer function, then performing the summation over all frequency components gives the time domain relation

$$\dot{w}(t) = \frac{1}{8\sqrt{\rho h D}}F(t) \tag{6.56}$$

That is, the contact force is directly related to the particle velocity at the impact site. Compare this with the equivalent hereditary relationship determined for the beam. This is useful in obtaining the impacting force history as demonstrated in Appendix A. Reference [152] shows some applications of this formula.

Strain Histories

The radial and hoop strains are obtained by differentiation of the displacement to give

$$\hat{e}_{rr} = \frac{i\hat{F}h}{32D}\{(J_0 - J_2) - i(Y_0 - Y_2) + \frac{i2}{\pi}(K_0 + K_2)\}$$

$$\hat{e}_{\theta\theta} = \frac{i\hat{F}h}{32D}\{(J_0 + J_2) - i(Y_0 + Y_2) + \frac{i2}{\pi}(K_0 - K_2)\} \qquad (6.57)$$

This solution has a peculiarity in that at $r = 0$ the strain is infinite (even though the displacements are finite) and given by

$$e_{rr} = e_{\theta\theta} = \frac{iFh}{32D}\left[1 - \frac{4i}{\pi}\ln(r)\right]$$

A thick-plate solution could be used to eliminate this singularity. But it can be done more directly by realizing that a strain gage gives the value of strain averaged over its length. That is, if the gage is of length L and centered at $r = 0$, then its strain reading is

$$e_{rr} = \frac{iFh}{32D}\left[1 - \frac{4i}{\pi}\ln\left(\frac{\beta L}{2e}\right)\right]e^{i\omega t}$$

This is nonsingular for all finite L. When strain measurements are made at the impact site, it is this equation that is used in the analysis.

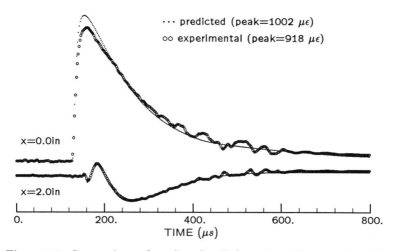

Figure 6.6: Comparison of predicted radial strains with experimental.

Sample strain histories are shown in Figure 6.6. The force history used was obtained from the impact of an aluminum plate with a steel sphere and using the

Hertzian contact law. The equations used are given in Appendix C. The strain histories have some very interesting features. First, the peak strain observed decreases very rapidly with position. Secondly, they all tend to the same limiting behavior which can be shown to be proportional to

$$\frac{1}{t}\cos[\beta^2 r^2/(4t)]$$

The strain decreases in time much more rapidly than for a beam which has a $1/\sqrt{t}$ dependence.

For comparison, some experimentally recorded strain histories are also given. The agreement is very good, although the theory slightly overestimates the strain at the impact site. Since these experimental strains will be used in the example on force identification their origin will be briefly summarized. The basic experimental setup is detailed in Reference [65]. An aluminum plate 0.065 inches thick was used. The Norland® 3001 digitally recorded the data every microsecond $(1\mu s)$ for a total of 4096 points per channel.

A steel ball (0.5 inch in diameter) is used as the impactor. A 1.5-V battery connected to it and the plate provided a voltage signal whose trace gives an idea of the contact time. The gages indicate significant reflections while the contact voltage trace indicates multiple impacts. The traces of Figure 6.6 has most of the reflections removed.

Once the data are collected, they are transferred in digital form to the computer where it is decoded and scaled. It is also smoothed by use of a three-point moving average. This constitutes the raw data used for pre-processing.

Force Identification

If only the radial strain is considered, then equation (6.57) can be written in the following ways

$$\hat{e}_n(r_1) = \hat{F}_n \bullet H_n(r_1) \tag{6.58}$$

$$\hat{F}_n = \hat{e}_n(r_1)/H_n(r_1) \tag{6.59}$$

$$\hat{e}_n(r_2) = \hat{e}_n(r_1) \bullet H_n(r_2)/H_n(r_1) \tag{6.60}$$

This corresponds, respectively, to the three cases: (1) given the force obtain the strain at r_1, (2) given the strain at r_1, obtain the force, and (3) given the strain at r_1 obtain the strain at r_2. In each case the force is obtained at the center of the plate $(r = 0)$. Of course, once the new FFT components are known the time histories are obtained simply by using the FFT inverse.

The basic scheme to obtain the force is to take the measured strain history $e(t)$, sample it every ΔT, obtain the Fourier components $\hat{e}_n$ and use equation (6.59) to obtain $\hat{F}_n$. However, since both the plate and sample window are finite, it is necessary to do some pre-processing of the strain data. The three main operations as discussed in Chapter 2 are: removing obvious reflections, appending the signal with its long-term behavior, and padding with zeros. Appending the signal improves the estimate of low-frequency content. The amount of padding needed depends on the distance of the gage from the impact site. An estimate of the amount needed is given by

$$t_{\text{pad}} = x_{\text{gage}} \sqrt{\frac{N \Delta T}{8\pi}} \left[\frac{\rho h}{D} \right]^{1/4}$$

For the above aluminum plate a gage located at 2in. (50mm) would require 250 μs padding for a window of 1000 μs, but 500 μs padding for a window 5000 μs. It will be shown later that smoothing out reflections can be the most challenging part of the pre-processing of the data.

In the present case there is ambiguity between some of the reflections and the multiple impacts. The data are first given a five-point average smoothing. The data from 600 μs onward are then given an 801-point average smoothing. The resultant processed signals are the ones shown in Figure 6.6.

The motivation for the five-point average is that the data will later be sampled by the FFT every 2 μs and this ensures that there are no aliasing effects. The purpose of the 801-point average is to remove reflections. This type of scheme can be successful because the mean of the reflections is zero. However, the period of the reflections changes thus requiring a variable-point average. This is accomplished in the present studies by the following: The first point filtered is given a three-point average, the next a five-point average and so on up to the maximum where it is kept constant. The raw signal is padded with zeroes to give a smooth transition to zero at the trailing edge of the processed data.

A 1024-point FFT is used with a sampling time of 2 μs. Since the reflections have different phase relationships at each gage location, then combining them can be used to further smooth out the reflections. If two gages are used for the force estimation, then this is best done by estimating the spectra at a common location and averaging.

Figure 6.7 shows the reconstructions of the force history. They show good agreement with each other, the largest differences being where the reflections are significant. The predicted contact times are close to those obtained from the contact voltage. Also shown in this figure is the force history predicted using the Hertzian contact law. The agreement is also very satisfactory, indicating that the Hertzian contact is good for steel on aluminum.

These results confirm the basic validity and accuracy of the spectral approach to force identification. They also show that the scheme to remove the singularity in the strain is correct. There are two main areas of sensitivity. First, the further away from the impact site the gage is located, the weaker the signal is and consequently the greater the noise level in the predicted force. Secondly, reflections tend to obscure the estimated long-term behavior of the strains thus giving poorer force

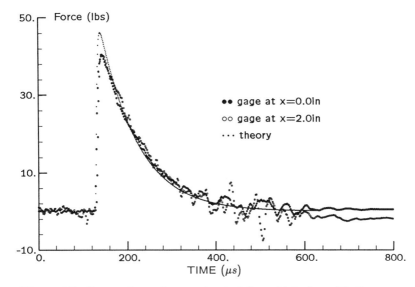

Figure 6.7: Comparison of reconstructed force histories with theory.

estimates when the contact time is relatively long. There are a few aspects of the method that need further study. Since the force history for plates usually has a very short rise time followed by a long trailing edge, this means that both high- and low-frequency components are important. This translates into using a large N for the FFT. If, however, an N of 1024 is set as the maximum, then it would be of interest to investigate the possibility of dividing the gage signal into high- and low-frequency components, processing and propagating them separately, and then reconstructing.

While the effects of reflections can be minimized by using a very large plate, this often is inconvenient or expensive. It would therefore be of interest to also investigate ways of modifying the algorithm so as to automatically remove reflections. This can be done using the more complicated solution of the following section for reflections.

6.7 Double Series Solution for Arbitrary Waves

A previous section introduced the application of the spectral methodology to the analysis of plane waves incident on straight boundaries. The problem is considerably more complicated if the wave and the boundaries are described by different coordinate system. This is the focus of this section.

To make the treatment realistic, the incident wave (initiated through point impact with a steel ball) having circularly crested wavefronts will be taken to interact with a rectilinear boundary. Thus, the first task is to extend the spectral method so as to handle spatial variations in two dimensions. The adequacy of the double summation scheme introduced is first verified by reanalyzing the point impact problem treated in the last section. Some of the factors that go into making a good 2-D spectral analysis are discussed. It is then applied to the analysis of the boundary problems. An interesting aspect of this is the possibility of generating surface waves analogous to Rayleigh waves. It is shown, however, that these are dispersive and so are not as discernible or damaging as their non-dispersive counterpart.

Double Series Solution

The approach taken in this and the next few sections is basically that of Lamb's 1904 classic paper on seismic tremors (Reference [133]). There are two significant differences, however. The first is that the waves obey the flexural dispersion relations. The second is that whenever the time domain behavior is required the FFT computer algorithm is used for the reconstructions. The functions $\hat{w}_1$ and $\hat{w}_2$ are analogous to the P-wave and S-wave potentials in Lamb's development.

Consider solutions of the form

$$\hat{w}_1 = \mathbf{A}e^{-ikz}\cos\eta y\,, \qquad \hat{w}_2 = \mathbf{B}e^{-\bar{k}z}\cos\eta y \tag{6.61}$$

where k, $\bar{k}$ and η are wave numbers. This form (which gives behaviors symmetric about the $y = 0$ line) is adequate for the present purpose although the more general $e^{-i\eta y}$ could also easily be used. The characteristic equations associated with these solutions are

$$-k^2 - \eta^2 + \beta^2 = 0\,, \qquad +k^2 - \eta^2 - \beta^2 = 0$$

giving the spectrum relations as

$$k = \pm\sqrt{\beta^2 - \eta^2}\,, \qquad \bar{k} = \pm\sqrt{\beta^2 + \eta^2} \tag{6.62}$$

The solution for the plate becomes

$$w(x,y,t) = \sum\sum[\mathbf{A}e^{-ikz} + \mathbf{B}e^{-\bar{k}z} + \mathbf{C}e^{+ikz} + \mathbf{D}e^{+\bar{k}z}]\cos\eta y\, e^{i\omega t} \tag{6.63}$$

That is, the actual solution is obtained by summing kernel solutions of the above form for many values of ω and η. In elementary terms, it can be thought of as containing a forward-moving wave $\mathbf{A}$, a backward-moving wave $\mathbf{C}$ and corresponding evanescent waves $\mathbf{B}$ and $\mathbf{D}$. However, because the wavenumber k can be complex under some circumstances, then the above distinctions are sometimes blurred. To amplify this idea, consider the solution as a plane wave in x (the bracketed term) modified in y. Then, for a particular η, the summation over ω is similar to that for a beam in the last chapter. The corresponding spectrum relations, however, are modified by η as shown in Figure 6.8. It is noted that for a particular η the first mode shows a cut-off frequency with the lower-frequency components being non-propagating.

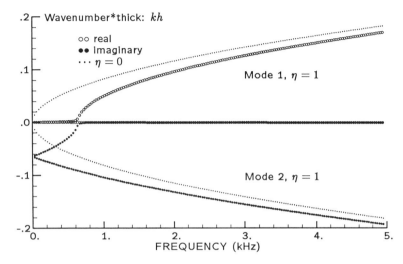

Figure 6.8: The spectrum relation for different values of the y wavenumber.

The present formulation is set up to solve problems whose boundary conditions are specified along $x =$ constant. At the boundary $x = R$, the solution and the four conditions can be written as

$$
\begin{aligned}
\hat{w}(x,y) &= \left[\mathbf{A}e^{-ikx} + \mathbf{B}e^{-\bar{k}x} + \mathbf{C}e^{+ikz(z-R)} + \mathbf{D}e^{\bar{k}(z-R)}\right]\cos\eta y \\
\hat{w} &= [\mathbf{A}^* + \mathbf{B}^* + \mathbf{C} + \mathbf{D}]\cos\eta y \\
\frac{\partial\hat{w}}{\partial x} &= [-ik\mathbf{A}^* - \bar{k}\mathbf{B}^* + ik\mathbf{C} + \bar{k}\mathbf{D}]\cos\eta y \\
\frac{1}{D}\hat{M}_x &= [-(k^2 + \nu\eta^2)\mathbf{A}^* + (k^2 - \nu\eta^2)\mathbf{B}^* - (k^2 + \nu\eta^2)\mathbf{C} + (\bar{k}^2 - \nu\eta^2)\mathbf{D}]\cos\eta y \\
\frac{-1}{D}\hat{V}_x &= [ik(\beta^2 + (1-\nu)\eta^2)\mathbf{A}^* - \bar{k}(\beta^2 - (1-\nu)\eta^2)\mathbf{B}^* \\
&\qquad -ik(\beta^2 + (1-\nu)\eta^2)\mathbf{C} + \bar{k}(\beta^2 - (1-\nu)\eta^2)\mathbf{D}]\cos\eta y \qquad (6.64)
\end{aligned}
$$

where the abbreviations $\mathbf{A}^* = \mathbf{A}e^{-ikR}$, $\mathbf{B}^* = \mathbf{B}e^{-\bar{k}R}$ have been used. The particular problems of interest will be constructed by combinations of these.

Point Impact Response of a Large Plate

This problem is used to validate the implementation of the double summation approach as well as to establish the incident wave conditions for the edge problems. The outward propagating solution in the region $x > 0$ is

$$\hat{w} = [\mathbf{A}e^{-ikz} + \mathbf{B}e^{-\bar{k}z}]\cos\eta y \tag{6.65}$$

Since this solution must be symmetric about $x = 0$, then the slope with respect to x is zero, giving $\bar{k}\mathbf{B} = -ik\mathbf{A}$. The shear near $x = 0$ is from equation (6.64)

$$\frac{-1}{D}V_z = \mathbf{A}ik2\beta^2\cos\eta y$$

Let the applied force spectrum be represented by the Fourier cosine series

$$p(y) = \sum_m a_m\epsilon_m\cos\eta_m y = \frac{\hat{F}}{L}\sum_m \epsilon_m\cos\eta_m y$$

$$\epsilon_o = 0.5, \qquad \epsilon_m = 1, \qquad \eta_m = 2\pi m/L \tag{6.66}$$

where the second form is specifically for a concentrated load and L is the space window. The coefficient $\mathbf{A}$ can now be determined giving the deflected shape as

$$w(x,y,t) = \sum_n \frac{-i\hat{F}}{8D\beta^2}\frac{4}{L}\sum_m \epsilon_m[\frac{e^{-ikz}}{k} - i\frac{e^{-\bar{k}z}}{\bar{k}}]\cos\eta y\, e^{i\omega t} \tag{6.67}$$

It is interesting to note that if $\eta = 0$, then $k = \bar{k} = \beta$ and the solution for the impact of a beam is recovered. Also of interest is the comparison with the solution using Bessel functions given in equation (6.55). It is seen that both solutions are the same since the Bessel functions in polar coordinates have the cartesian representation

$$[J_o - iY_o - \frac{i2}{\pi}K_o] \Rightarrow \frac{4}{L}\sum_m[\frac{e^{-ikz}}{k} - i\frac{e^{-\bar{k}z}}{\bar{k}}]\cos\eta y$$

This is analogous to Lamb's (1904) relation (13) except that here the summation is finite.

Figure 6.9 shows the comparison of the experimental strains of Figure 6.6 with those predicted by the above analysis. The results establish the equivalence of the Bessel and double series solutions. The specifications for the computations are

$$\begin{array}{ll} \text{Time window} = 2048 \ \mu s & N = 1024 \\ \text{Space window} = 128 \text{ ins } (3.25 \text{ m}) & M = 256 \end{array}$$

Since the wavenumber k in the denominator can be zero, a small amount of damping is added to the system by making the substitution

$$\beta \qquad \Rightarrow \qquad \beta(1 - 0.01\omega)$$

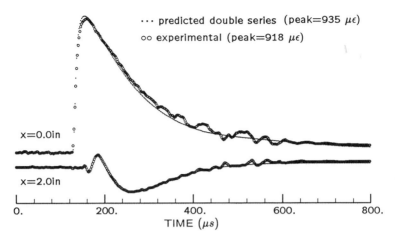

Figure 6.9: Comparison of experimental and predicted strains.

This artifice also removes the ambiguity when obtaining square roots as in equation (6.62).

Although both solutions are equivalent, the one containing the Bessel functions is preferable because then only a single series is involved and its summation is accomplished very economically using the FFT computer algorithm. In fact, on an 8 MHz IBM-AT® style microcomputer a Bessel's solution will take on the order of seconds but the comparable double series solution takes twenty minutes. Unfortunately, the Bessel form of solution is not suitable for imposing boundary conditions along a straight edge and so the double series form must be used. If the response is required at many instances of time and many locations in space, then a double FFT algorithm should be used for the inversion. In the present circumstances, however, the solution is required continuous in time but only at a limited number of space locations (because the experimental comparison is with strain gages and that is the nature of strain gage data). Consequently, the summations over m are performed longform, but those over n are done using the FFT.

Waves Reflected from a Straight Edge

Reference [117] considers, quite extensively, the reflection of straight-crested flexural waves from the edge of a semi-infinite plate. Incident circular-crested waves, however, are a different matter because in order to satisfy the boundary conditions the reflected waves must behave more complicatedly. By analogy with the in-plane case earlier in this chapter, Rayleigh surface waves must be generated in order to maintain the zero traction condition. Three typical types of boundaries are used here so as to illustrate the nature of the reflected waves. The relations are obtained by taking special forms of equation (6.64) as needed.

A simply supported edge condition requires that at the boundary

$$w = 0, \qquad M_x = 0$$

giving the simple result

$$\mathbf{C} = -\mathbf{A}e^{-ikR}, \qquad \mathbf{D} = -\mathbf{B}e^{-\bar{k}R} \tag{6.68}$$

and in full form

$$\hat{w}_{\text{ref}} = -[\mathbf{A}e^{ik(z-2R)} + \mathbf{B}e^{\bar{k}(z-2R)}]\cos \eta y \tag{6.69}$$

Thus, the wave appears to originate from an image source a distance $2R$ away. However, this is the only case where the imaging idea applies; in general, the new waves will originate at the boundary itself.

For a fixed edge condition

$$w = 0, \qquad \frac{\partial w}{\partial x} = 0$$

giving as the system to be solved

$$\begin{bmatrix} 1 & 1 \\ ik & \bar{k} \end{bmatrix} \begin{Bmatrix} \mathbf{C} \\ \mathbf{D} \end{Bmatrix} = \begin{bmatrix} -1 & -1 \\ ik & \bar{k} \end{bmatrix} \begin{Bmatrix} \mathbf{A}^* \\ \mathbf{B}^* \end{Bmatrix}$$

Using Cramer's Rule this gives

$$\mathbf{C} = \{(ik + \bar{k})\mathbf{A}e^{-ikR} + 2\bar{k}\mathbf{B}e^{-\bar{k}R}\}/\Delta$$
$$\mathbf{D} = -\{2ik\mathbf{A}e^{-ikR} + (ik + \bar{k})\mathbf{B}e^{-\bar{k}R}\}/\Delta$$
$$\Delta = ik - \bar{k} \tag{6.70}$$

The magnitude of the determinant is related to

$$\Delta^2 = (ik - \bar{k})(-ik - \bar{k}) = k^2 + \bar{k}^2 = 2\beta^2$$

and is never zero. The presence of both e^{-ikR} and $e^{-\bar{k}R}$ in both C and D shows that the reflected waves cannot be written simply in terms of image waves.

A free edge boundary has the following two conditions

$$M_x = 0, \qquad V_x = 0$$

The system of equations to be solved is

$$\begin{bmatrix} k_\nu^2 & -\bar{k}_\nu^2 \\ ik\bar{k}_\nu^2 & -\bar{k}k_\nu^2 \end{bmatrix} \begin{Bmatrix} \mathbf{C} \\ \mathbf{D} \end{Bmatrix} = \begin{bmatrix} -k_\nu^2 & \bar{k}_\nu^2 \\ ik\bar{k}_\nu^2 & -\bar{k}k_\nu^2 \end{bmatrix} \begin{Bmatrix} \mathbf{A}^* \\ \mathbf{B}^* \end{Bmatrix}$$

where

$$k_\nu^2 = \beta^2 - (1 - \nu)\eta^2, \qquad \bar{k}_\nu^2 = \beta^2 + (1 - \nu)\eta^2$$

The solution is

$$\mathbf{C} = \{(ikk_\nu^4 + \bar{k}k_\nu^4)\mathbf{A}e^{-ikR} - 2\bar{k}\bar{k}_\nu^2 k_\nu^2 \mathbf{B}e^{-\bar{k}R}\}/\Delta$$
$$\mathbf{D} = \{2ikk_\nu^2 \bar{k}_\nu^2 \mathbf{A}e^{-ikR} - (ik\bar{k}_\nu^4 + \bar{k}k_\nu^4)\mathbf{B}e^{-\bar{k}R}\}/\Delta$$
$$\Delta = ik\bar{k}_\nu^4 - \bar{k}k_\nu^4 \tag{6.71}$$

It is apparent that the coefficients **C** and **D** have a very complicated dependence on both frequency w and wavenumber η. With the present spectral formulation this actually poses no problems when obtaining the inverses (or time reconstructions).

Reference [68] shows a comparison of the predictions of the free edge strains with those measured at a few selected points. The modeling does a very good job in capturing the essence of the experimental behavior. The biggest deviations occur toward the trailing end but this is the area where the force is not known exactly because (as mentioned previously) reflections interfere with the dynamics of the ball. Also, the experimental values are not known with confidence because of the presence of reflections in the recorded traces.

The Possibility of Rayleigh-Type Edge Waves

The determinant of equation (6.71) can be zero, therefore raising the possibility of a singular solution for [**C, D**]. In its simplest form, it is of interest to know if the wave

$$w = [\mathbf{C}e^{ikz} + \mathbf{D}e^{\bar{k}z}]e^{i\eta y}e^{i\omega t} \tag{6.72}$$

can propagate along the edge and satisfy the free edge conditions. From equations (6.64) this gives

$$\begin{bmatrix} k_\nu^2 & -\bar{k}_\nu^2 \\ ik\bar{k}_\nu^2 & -kk_\nu^2 \end{bmatrix} \begin{Bmatrix} \mathbf{C} \\ \mathbf{D} \end{Bmatrix} = 0$$

This is possible only if the determinant

$$\Delta = ik\bar{k}_\nu^4 - \bar{k}k_\nu^4 = 0 \tag{6.73}$$

is zero. (This expression for the determinant is obviously the same as for the last section.) The determinant behavior (in the form of its reciprocal) is shown in Figure 6.10. It is apparent that the determinant can go through zero and this indicates the possibility of a new edge wave being formed analogous to Rayleigh surface waves. However, the present edge wave is dispersive and so it will not be so readily observable. The wavenumbers where this happens are (after expanding equation (6.73))

$$\eta = \left[\frac{(1 - 3\nu) + 2\sqrt{1 - 2\nu + 2\nu^2}}{(1 - \nu)^2(3 + \nu)}\right]^{1/4} \beta, \qquad k = \sqrt{\beta^2 - \eta^2} \tag{6.74}$$

The quantity inside the bracket is close to unity for all values of the Poisson's ratio. For example

$$\nu = 0.25, \quad \eta = 1.00041\beta, \quad k = 0.0287i\beta$$
$$\nu = 0.30, \quad \eta = 1.00095\beta, \quad k = 0.0436i\beta$$
$$\nu = 0.35, \quad \eta = 1.00196\beta, \quad k = 0.0626i\beta$$

Thus the edge wave travels at approximately the same speed as the incident wave and is attenuated into the plate. This wave is difficult to isolate because it is dispersive (since β is dispersive) and it arrives later than the corresponding incident wave that generated it.

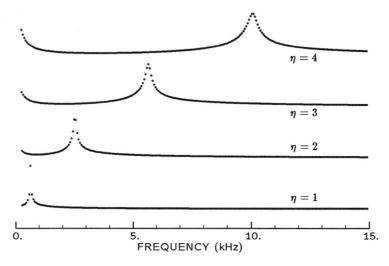

Figure 6.10: Determinant behavior for different values of the y wavenumber.

6.8   Anisotropic Plates

As is apparent from the previous chapters on rods and beams, the extension to more complicated plates is quite straightforward; it generally is a matter of just changing the spectrum relation. The new generalization of relevance to plates is that of anisotropic material behavior. That is, the plates can have different properties in different directions. This is of significance today because of the greater use of fiber-reinforced materials. A problem of relevance to composite materials is that of obtaining its mechanical properties. The application of spectral analysis to doing this is demonstrated.

The mechanics of composite materials is covered in depth in References [9,113], so only the relevant equations will be stated here.

Spectral Analysis

When the stress strain relation is anisotropic, then the resultants also exhibit anisotropy. Thus the moment-deflection relation becomes

$$
\left\{
\begin{array}{c}
M_x \\
M_y \\
M_{xy}
\end{array}
\right\}
=
\left[
\begin{array}{ccc}
D_{11} & D_{12} & D_{16} \\
D_{12} & D_{22} & D_{26} \\
D_{16} & D_{26} & D_{66}
\end{array}
\right]
\left\{
\begin{array}{c}
\kappa_x \\
\kappa_y \\
\kappa_{xy}
\end{array}
\right\}
\tag{6.75}
$$

where the curvatures are related to the deflection by

$$
\kappa_x = \frac{\partial^2 w}{\partial x^2}, \qquad \kappa_x = \frac{\partial^2 w}{\partial y^2}, \qquad \kappa_{xy} = \frac{\partial^2 w}{\partial x \partial y}
$$

and the elements D_{ij} are the anisotropic stiffnesses. The equation of motion then becomes

$$
D_{11}\frac{\partial^2 w}{\partial x^4} + 2(D_{11} + 2D_{66})\frac{\partial^4 w}{\partial x^2 \partial y^2} + D_{22}\frac{\partial^4 w}{\partial y^4}
$$

$$
+ 2D_{16}\frac{\partial^4 w}{\partial x^3 \partial y} + 2D_{26}\frac{\partial^4 w}{\partial x \partial y^3} = -\rho h \frac{\partial^2 w}{\partial t^2}
\tag{6.76}
$$

Consider a plane wave propagating in an arbitrary direction θ

$$
w = \sum \mathbf{A} e^{-i[k(xC_\theta + yS_\theta)]} e^{-i\omega t}
$$

then the characteristic equation is

$$
k^4[D_{11}C_\theta^4 + 2(D_{11} + 2D_{66})\, C_\theta^2 S_\theta^2 + D_{22}S_\theta^4 + 2D_{16}C_\theta^3 S_\theta + 2D_{26}C_\theta S_\theta^3] = \rho h \omega^2
$$

This gives the spectrum relation as

$$
k = \sqrt{\omega}\left[\frac{\rho h}{D_{\theta\theta}}\right]^{\frac{1}{4}}
$$

$$
D_{\theta\theta} = D_{11}C_\theta^4 + 2(D_{12} + 2D_{66})C_\theta^2 S_\theta^2 + D_{22}S_\theta^4 + 2D_{16}C_\theta^3 S_\theta + 2D_{26}C_\theta S_\theta^3
\tag{6.77}
$$

Formally, it is the same as for the isotropic plate except that the stiffness appearing in the relation is a function of the orientation of the wave.

Material Properties Testing of Composites

Many materials exhibit different mechanical properties dynamically than statically, so it is of interest to be able to measure properties dynamically. This section shows an example of the in-situ testing of the flexural properties of a composite laminate. The basis of the method is to take two gage measurements and relate the phase changes of the wave to the stiffness properties. This is possible since it has just been shown that the material properties are embedded in the wavenumber by

$$k = \sqrt{\omega} \left[\frac{\rho h}{D_{\theta\theta}} \right]^{1/4} = \beta \tag{6.78}$$

This is a brief summary of the work reported in Reference [71]. The Bessel function form for the impact of plates is rewritten so as to make β easy to extract

$$\hat{e}_r = \sum \beta 2 \sqrt{\frac{2}{\pi \beta r}} e^{-i\beta r} [Q] e^{i\omega t} \tag{6.79}$$

where Q is a complex function generally of unity magnitude given by

$$Q(r,\omega) \equiv \sqrt{\frac{\pi \beta r}{8}} e^{i\beta r} \left[(J_0 - J_2) - i(Y_0 - Y_2) + \frac{i2}{\pi}(K_0 + K_2) \right]$$

Although this solution is based on the impact of isotropic plates, Reference [72] shows its reasonable use for orthotropic plates. Let the time trace of each strain gage be represented in its spectral form as $\hat{e}_r(r, \omega_n)$, then the following association can be made

$$\hat{e}_r(r, \omega_n) = \beta_n 2 \sqrt{\frac{2}{\pi \beta_n r}} e^{-i\beta_n r} [Q_n]$$

The wavenumber is now simply obtained from the two measurements as the argument of the complex function

$$\beta_n = \frac{1}{(r_1 - r_2)} \text{ argument } [\hat{e}_r(r_1, \omega_n) \hat{e}_r^*(r_2, \omega_n) / (Q_{n1} Q_{n2}^*)]$$

In this expression the superscript asterisk means complex conjugate. The argument is obtained by taking the arctangent of the imaginary part of the function over the real part. Unfortunately, this is known only to within multiples of π radians. Since it is essential to have the correct absolute value of phase, the following simple method is used:

$$\text{Phase} = \arctan[e^{-i\beta x}] + \beta x$$

This ruse (which was also used in the section on remote sensing) basically propagates the function back to where the variation of phase is within π radians, the arctangent is then taken and finally the subtracted phase is replaced as the $+\beta x$ term. The scheme works because the function form of β is known and good estimates for the parameters can be made. The value of x can be adjusted so that discontinuities are not observed in the phase versus frequency plot.

In the above expression the complex function $Q_1 Q_2^*$ contains the unknown β. However, its effect on the phase is quite small for monitoring sites greater than 2 inches and is noticeable only at low frequencies. Thus, if a combination of large r

and high frequencies is used, then it would be reasonable to neglect Q altogether. If this is not feasible, then its effect can be accounted for by guessing a value of $D_{\theta\theta}$, computing Q and then β. This in turn gives a better guess for $D_{\theta\theta}$ and the process is continued. In practice no more than one iteration is usually needed to account for Q exactly.

Reference [71] shows plots of the experimentally determined wavenumber as a function of frequency. A further extension of the method is shown in Reference [62] where non-contacting fiber optics are used as the sensors. This is a practical scheme because only the phase relations are required; thus it is unnecessary to calibrate the optical sensors.

Other Solutions

As a final point, it is possible to obtain exact solutions for the impact of an orthotropic plate by beginning with a kernel solution of the form

$$\hat{w}(x, y) = e^{i\eta y} e^{-ikx}$$

and substituting in equation (6.76). In fact, the series approach will allow the solution of many problems in an orthotropic plate. Their details, however, are beyond the scope of this book.

Afterword

".... one moves forward from this mystic landing to Fourier analysis, periodogram analysis, Fourier analysis over groups, and one comes rapidly to great technological applications, and always a sense of the actual and potential unities that lurk in the corners of the universe."

P. J. DAVIS and R. HERSH — The Mathematical Experience
Birkhäuser Boston, 1981

It is opportune now to assess the accomplishments of this book and (perhaps more important) to point out what is missing.

This book has emphasized the analysis of structures in situations that facilitate their treatment as connected waveguides. Thus, the primary effort is devoted to developing the spectral analysis method for solving wave propagation problems in rods, beams and plates. An introduction to matrix methods and to the higher-order structural theories extends the range of application of these models to general structures. The numerous examples place special emphasis on the waves interacting with structural discontinuities. It is shown that within the spectral framework, the incorporation of damping, coupling effects and higher-order theories, is a straightforward and simple matter. Additionally, the incorporation of experimental data and results in many of the examples demonstrates the affinity of the spectral methods for experimental methods.

What emerges is an approach that exhibits a striking unity between the analytical, the computational and the experimental methodologies.

A topic introduced but not fully developed is represented by the truss impact problem of Chapter 5. The matrix method approach demonstrated is very exciting because it offers the possibility of doing wave analysis in complicated structures without having to resort to expensive, time-consuming cascading methods. The efficiency of the approach resides in the fact that the only unknowns are those associated with the connectivities at the joints and therefore are few in number. The beauty of such an approach is that the structure can be analyzed as a whole without worry of the detailed histories. (Only if necessary is the detailed wave behavior reconstructed, and this can be done quite efficiently as part of the post-processing.) A further side benefit is that it brings wave propagation and vibration analyses closer together and it makes such global methods as modal analysis possible, but all the time it still has the possibility of recovering the detailed behavior. But it seems that what is really needed in structural dynamics analyses is some global or higher level evaluation of the results. The spectral analysis approach has the potential to offer this, but it is not developed as yet.

Significant by their absence is the whole series of problems dealing with waves in extended media interacting with discontinuities (such as holes or cracks) causing dynamic stress concentrations, diffraction, and scattering. These problems can be treated by spectral analysis but they require the development of some new tools as well as the use of double series (an approach briefly introduced in Chapter 6). These topics are really part of the more general one of the exact solution of waves in waveguides. While many such exact solutions can be found in the literature, they generally are too cumbersome to be of direct value. To date, the only research done is either to document the spectrum relation or to justify the lower order approximate solutions. The use of the FFT offers the possibility for developing an effective, efficient tool for doing this analysis but simpler schemes for determining the spectrum relation for multi-mode situations must first be developed. Additionally, the apparatus for efficiently utilizing the double FFT must also be put in place.

These two problems represent different departures from the contents of this book. The first is to a level of integration, the second to a level of more detail. It is hoped that the methodologies presented thus far has provided a useful middle ground for beginning both studies.

Appendix A

Contact Force

Generally, the wave motion in a structure is due to some impact or blast loading. To make the analyses of this book practical, it is therefore useful to have some way of estimating the force histories resulting from this. Following is a collection of results that allow force estimates for the rod, beam and plate, and they are coded in the program FORCCOMP in Appendix D.

The analysis involves three aspects: the motion of the structure, the motion of the impactor, and then the contact relation between the structure and impactor. The appropriate equations for each of these are summarized.

Equations of Motion for the Structure

While it is obvious that the equations of motion of the structure can be written in terms of partial differential equations, the need here is to be able to write a relation between the applied force and the structural response at the point of impact.

The approach used for the rod is based on an idea used in References [192,17]. Consider a long bar impacted at one end and assume that a longitudinal plane wave is generated. The resulting velocity of an arbitrary point is related to the impacting force (as shown in Chapter 3) by

$$\dot{u}_b = -\frac{c_o}{EA}F \tag{A.1}$$

Now let it be assumed that this relation is valid even near the impact site.

The description of the response of the beam in the time domain is not as simple as for the rod and many approaches have been taken as can be seen from References [24,134,166]. The approach here is to use a result from Chapter 4 where it was shown that the equations of motion of the beam at the impact site are

$$\dot{v}(o,t) = \frac{2}{h}\sqrt{\frac{EI}{\rho A}}B\int_o^t \frac{F(\tau)}{\sqrt{t-\tau}}d\tau, \qquad B = \frac{h}{4EI\sqrt{2\pi}}\left[\frac{EI}{\rho A}\right]^{\frac{1}{4}} \tag{A.2}$$

Since this relation is a singular integral numerical methods must be developed in order to utilize it. In fact, Reference [56] shows how this can be done even when x is not zero (i.e., when the observation point is not the impact point).

Let it be assumed that the response of an impacted plate can be represented adequately by classical plate theory. As shown in Chapter 6, the velocity and force at the impact site are related by

$$\dot{w}(t) = \frac{1}{8\sqrt{\rho h D}} F(t) \tag{A.3}$$

That is, the contact force is directly related to the particle velocity at the impact site.

Equation of Motion for the Impactor

Assume that the impacting mass can be considered concentrated, then its equation of motion is

$$m_s \ddot{u}_s = -F \tag{A.4}$$

If the impactor itself is a structure of some extent (a rod, for instance), then one of the above equations of motion can be used.

Contact Laws

The simplest contact law is that of plastic impact. Here it is assumed that after contact is first made, the contacting particles of the mass and structure move together. That is, for the rod

$$\dot{u}_s = \dot{u}_b \tag{A.5}$$

Ther are similar relations for the beam and plate. In the case of the rod and plate, this law allows the equation of motion to be rewritten as

$$m_s \ddot{u}_s = -F = -\frac{EA}{c_o}\dot{u}_b = -\frac{EA}{c_o}\dot{u}_s$$

This gives uncoupled differential equations for the mass and rod as

$$\frac{d}{dt}[\dot{u}_s] + \frac{EA}{m_s c_o}[\dot{u}_s] = 0, \qquad \frac{d}{dt}[\dot{u}_b] + \frac{EA}{m_s c_o}[\dot{u}_b] = 0$$

These identical relations can be integrated to give

$$\dot{u}_b = \dot{u}_s = V_s e^{-t/\tau}, \qquad \tau = \frac{m_s c_o}{EA}$$

where the initial condition of $\dot{u}_s = V_s$ has been imposed. The generated force history in the rod is then

$$F = V_s \frac{EA}{c_o} e^{-t/\tau} = -\frac{V_s m_s}{\tau} e^{-t/\tau} \tag{A.6}$$

This gives a very sharp rise in contact force followed by an exponential decay. The leading term in the second form can be viewed as the momentum of the impactor averaged over the characteristic time τ. Similar conclusions result for the beam and plate.

An entirely different contact law arises when the two contacting objects remain elastic. A classical solution for the contact of two elastic bodies is that due to Hertz (Reference [91] has a good review of this theory as well as some other types of contact laws). Because this theory takes into account the curvature of the contacting surfaces it results in a nonlinear contact law. The force between two spherical masses of radii R_1 and R_2 in contact is described by

$$F = k\alpha^{\frac{3}{2}} = k(u_1 - u_2)^{\frac{3}{2}} \tag{A.7}$$

where α is the relative indentation and

$$k = \frac{4}{3}\sqrt{\frac{R_1 R_2}{R_1 + R_2}}\left(\frac{k_1 k_2}{k_1 + k_2}\right), \qquad k_1 \equiv \frac{E_1}{1 - \nu_1^2}, \qquad k_2 \equiv \frac{E_2}{1 - \nu_2^2} \tag{A.8}$$

This is a nonlinear contact law. If the second body is flat, then $R_2 = \infty$ and if the first is very hard then $1/k_1$ is nearly zero giving, respectively,

$$k = \frac{4}{3}\sqrt{R_1}\left(\frac{k_1 k_2}{k_1 + k_2}\right), \qquad k = \frac{4}{3}\sqrt{R_1}\, E_2$$

In both cases the contact law still remains nonlinear. This nonlinearity makes obtaining closed form solutions very difficult.

The above results for the Hertzian contact, strictly speaking, are for two bodies in static contact, but it is usual to assume that for dynamic problems the static law still holds.

Numerical Integration

When the contact law is nonlinear, the equations must be integrated numerically. The rod will be used as the example to demonstrate how this is done, but the approach for the beam and plate are essentially the same.

For the rod let it be assumed that the one-dimensional relation is valid even near the impact site, then the indentation law can be written as

$$F = k\alpha^{\frac{3}{2}} = k(u_s - u_b)^{\frac{3}{2}}$$

and u_s and u_b are obtained from the following differential equations

$$\dot{u}_b = \frac{ck}{EA}(u_s - u_s)^{\frac{3}{2}} = \frac{ck}{EA}\alpha^{\frac{3}{2}}$$

$$\dot{V}_s = -\frac{k}{m_s}(u_s - u_b)^{\frac{3}{2}} = -\frac{k}{m_s}\alpha^{\frac{3}{2}}$$

$$\dot{u}_s = V_s$$

These are integrated subjected to the initial conditions

$$\text{at } t = 0: \qquad u_s = 0, \quad \dot{u}_s = V_s, \quad u_b = 0$$

Notice that the introduction of V_s as a dependent variable gives a system of three first-order ordinary differential equations. The code in the program FORCCOMP in Appendix D contains a scheme for their numerical integration based on the simple trapezoidal rule. A similar set of equations are set up for the plate. The beam requires to be treated a little differently because of the singular integral but this too is easily done numerically. Reference [17] reports some experimental results for impact with different size spheres on different diameters rods.

The situations treated in this appendix represent very special cases but they can be used to give good estimates of force histories in real situations. When more precise values are required, then using one of the experimentally measured contact laws should be considered. Another point to keep in mind is that since the spectral approach is capable of doing inverse problems, if some experimental data are available, the actual force history can be determined directly.

Appendix B

Bessel Functions

Bessel functions, and the other special functions, arise as solutions to many boundary value problems and their use can greatly extend the range of the spectral methods. The major properties of the Bessel functions are summarized here so as to facilitate their use. Also, a selection of useful references are given at the end of the appendix.

Bessel Equations and Solutions

An equation of the form

$$z^2 \frac{d^2 w}{dz^2} + z \frac{dw}{dz} + (z^2 - n^2) w = 0, \qquad n \geq 0$$

is called Bessel's equation and has the two independent solutions

$$w(z) = C_1 J_n(z) + C_2 Y_n(z)$$

where J_n, Y_n are Bessel functions of the first and second kind, respectively. A related Bessel equation (which can be obtained by replacing z with iz) is

$$z^2 \frac{d^2 w}{dz^2} + z \frac{dw}{dz} - (z^2 + n^2) w = 0$$

and has the two independent solutions

$$w(z) = C_3 I_n(z) + C_4 K_n(z)$$

where I_n, K_n are called modified Bessel functions of the first and second kind, respectively. When manipulating Bessel functions, the following recurrence relations are very useful:

$$M_{n+1} = \frac{2n}{z} M_n - M_{n-1}, \qquad M_{-1} = -M_1$$

$$-2\frac{dM_n}{dz} = M_{n+1} - M_{n-1}$$

$$L_{n+1} = -\frac{2n}{z}L_n + L_{n-1}, \qquad L_{-1} = +L_1$$

$$2\frac{dL_n}{dz} = L_{n+1} + L_{n-1}$$

where M_n represents any of the functions J_n, Y_n, H_n^1, H_n^2 and L_n either of $I_n, e^{in\pi}K_n$.

Many equations are transformable into Bessel equations and a very useful general form is

$$z^2\frac{d^2w}{dz^2} + (1+2\alpha)z\frac{dw}{dz} + (\beta^2 z^{2\gamma} + \delta^2)w = 0$$

This has the solution

$$w = \frac{1}{z^\alpha}[C_1 J_\nu(\frac{\beta z^\gamma}{\gamma}) + C_2 Y_\nu(\frac{\beta z^\gamma}{\gamma})], \qquad \nu = \sqrt{\alpha^2 - \delta^2}/\gamma$$

A similar general form can be set up for the modified functions.

Limiting Behavior

Bessel functions have the expansions

$$J_n(z) = (\tfrac{1}{2}z)^n \sum_{m=0}^{\infty} (-\tfrac{1}{4}z^2)^m \frac{1}{m!\Gamma(n+m+1)}$$

$$Y_n(z) = -\frac{1}{\pi}(\tfrac{1}{2}z)^{-n} \sum_{m=0}^{n-1} \frac{(n-m-1)!}{m!}(\tfrac{1}{4}z^2)^m + \frac{2}{\pi}\ln(\tfrac{1}{2}z)J_n(z)$$

$$-\frac{1}{\pi}(\tfrac{1}{2}z)^n \sum_{m=0}^{\infty} \{\psi(m+1) + \psi(n+m+1)\}(-\tfrac{1}{4}z^2)^m \frac{1}{m!(n+m)!}$$

$$I_n(z) = (\tfrac{1}{2}z)^n \sum_{m=0}^{\infty} (\tfrac{1}{4}z^2)^m \frac{1}{m!\Gamma(n+m+1)}$$

$$K_n(z) = \tfrac{1}{2}(\tfrac{1}{2}z)^{-n} \sum_{m=0}^{n-1} \frac{(n-m-1)!}{m!}(-\tfrac{1}{4}z^2)^m + (-1)^{n+1}\ln(\tfrac{1}{2}z)I_n(z)$$

$$+(-1)^n\tfrac{1}{2}(\tfrac{1}{2}z)^n \sum_{m=0}^{\infty} \{\psi(m+1) + \psi(n+m+1)\}(\tfrac{1}{4}z^2)^m \frac{1}{m!(n+m)!}$$

where Γ is the gamma function and $\psi(n) = \sum 1/n - .5772$. For small arguments, in particular,

$$J_0 \approx 1 - \tfrac{1}{4}z^2, \qquad Y_0 \approx \frac{2}{\pi}\{\ln(\tfrac{1}{2}z) + \gamma\}J_o + \frac{2}{\pi}\tfrac{1}{4}z^2$$

$$I_0 \approx 1 + \tfrac{1}{4}z^2, \qquad K_0 \approx -\{\ln(\tfrac{1}{2}z) + \gamma\}I_o + \tfrac{1}{4}z^2$$

It is apparent that the functions Y_0, I_0 exhibit singularities $z = 0$.

When the argument is increased, then these functions have the asymptotic behavior of

$$J_n \Rightarrow \sqrt{\frac{2}{\pi z}} \cos(z - \frac{\pi}{4} - \frac{n\pi}{2}), \quad Y_n \Rightarrow \sqrt{\frac{2}{\pi z}} \sin(z - \frac{\pi}{4} - \frac{n\pi}{2})$$

$$I_n \Rightarrow \frac{1}{\sqrt{2\pi z}} e^z, \quad K_n \Rightarrow \sqrt{\frac{\pi}{2z}} e^{-z}$$

J_n, Y_n exhibit a damped oscillation and I_n, K_n behave as exponentials. For this reason, the solutions are often written in terms of the Hankel functions of the first and second kind given by

$$H_n^1(z) \equiv J_n(z) + iY_n(z), \quad H_n^2(z) \equiv J_n(z) - iY_n(z)$$

These have the limiting behaviors

$$H_n^1 \Rightarrow \sqrt{\frac{2}{\pi z}} e^{i(z - \frac{\pi}{4} - \frac{n\pi}{2})}, \quad H_n^2 \Rightarrow \sqrt{\frac{2}{\pi z}} e^{-i(z - \frac{\pi}{4} - \frac{n\pi}{2})}$$

That is, they have the exponential form of the circular functions with the difference that they decrease in amplitude for large argument. Consequently, the use of the Bessel functions in spectral analysis of waves is similar to that of the circular functions.

Additional Reading

Below is a collection of books that have proven useful for understanding the special functions. The excellent reference book by Abramowitz and Stegun contains approximate numerical schemes for evaluating Bessel functions. The book by Press et. al. contains the actual FORTRAN-coding for doing this.

1. Abramowitz, M., and Stegun, I. A., 1965, *Handbook of Mathematical Functions*, Dover, New York.

2. Bowman, F., 1958, *Introduction to Bessel Functions*, Dover, New York.

3. Lebedev, N.N., 1972, *Special Functions and Their Applications*, Dover, New York.

4. McLachlan, N.W., 1947, *Theory and Application of Mathieu Functions*, Oxford Press, London.

5. Morse, P.M., and Feshbach, H., 1953, *Methods of Theoretical Physics*, Vol. I and II, McGraw-Hill, New York.

6. Press, W.H., Flannery, B.P., Teukolsky, S.A., and Vetterling, W.T., 1986, *Numerical Recipes*, Cambridge University Press, Cambridge.

7. Relton, F.E., 1965, *Applied Bessel Functions*, Dover, New York.

8. Stratton, J.A., Morse, P.M., Chu, L.J., and Hutner, R.A., 1941, *Elliptic Cylinder and Spheroidal Wave Functions*, Wiley Press, New York.

9. Tranter, C.J., 1968, *Bessel Functions*, Hart, New York.

10. Watson, G.N., 1962, *A Treatise on the Theory of Bessel Functions*, Cambridge Press, London.

Appendix C

Examples Parameters

While devising the examples in the text, a conscious effort was made to keep constant as many of the parameters as possible. To have repeated all this material in the text would have distracted from the points being made. This appendix attempts to collect together all the relevant parameters.

To help organize things, there is a standard rod, a standard beam, a standard pulse, and unless stated otherwise these are what were used.

```
ROD:        diameter=1.0 in.
            aluminum

BEAM:       width=1.0 in.
            thick=0.25 in.
            aluminum

ALUMINUM:   density=0.000247 lb/ci
            modulus=10.6E6 psi

PULSE:      0.000000   0
            0.001000   0
            0.001100   1000
            0.001300   0
            0.001500   0
            (secs)
```

Chapters 2 & 3

In all the examples the common transform parameters are

```
N = 512,     dT = 5e-6
```

For Figure 2.6 the frequency $f_o = 10000$ Hz. The five spring stiffnesses of Figure 3.5 are

> 1000,   50000,   100000,   500000,   10000000   psi

The joint length of Figure 3.8 is 50 inches, and the reduction in area is from 1.0 sq. in. to 0.25 sq. in. The viscoelastic material properties of Figure 3.10 are

> E1=0.5e6,   E2=0.1e6,   eta=0.3

Chapters 4 & 5

In all the examples the common transform parameters are

> N = 1024,   dT = 10e-6

For Figure 4.20, the thickness is 2.0 in. and width is 1.0 in. The shear modulus for the general cases was taken as the isotropic value of 4.0 msi. The distance propagated is 80 inches.

For Figure 5.11 and Figure 5.12, the length L is 40 inches, while the curvature of the bend is a radius of 20 in. The bend has the cross-sectional properties of the standard beam. The portal frame members of Figure 5.13 are of length 2.0 in. and are standard beams. The truss dimension of Figure 5.15 is $L = 5$ inches and the members are made from standard rods.

Appendix D

Source Code Listings

This appendix contains the code to many of the programs referred to throughout the text. They are written in Microsoft® FORTRAN version 4.0 but an attempt was made to make them as portable as possible by conforming to the ANSI 77 standard.

The original intent was to provide the full source code for all the programs in a fully documented form. This turns out to be impractical because of the number of pages required. It has, therefore, been necessary to cut back on the number of programs and reduce the amount of documentation. However, the programs that are listed are complete and not just program segments. The complete set of programs used in the text are available through the author or from *ikayex* SOFTWARE TOOLS, 615 Elston Road, Lafayette, Indiana 47905, USA.

The subroutines for CFFTCOMP are collected into major groups associated with their function which should help in navigating through the program.

CFFTCOMP

MAIN GROUP

```
c
c       main for CFFTCOMP
c
        integer size, dv0, dv1, dv2,dv3,dv4,dv5
c
        parameter( nmax =4100, size= 21000)
c
c       reserve master space and zero it
        real   respce(size)
        dimension x(nmax,2),wk(nmax,2),tt(nmax)
        equivalence (x(1,1),respce(1))
        equivalence (wk(1,1),respce(8201))
        equivalence (tt(1  ),respce(16401))
c
        do 2 n=1,size
           respce(n)=0.0
2       continue
```

```
c
      write (*,'(///////)')
      write(*,*)'CFFTCFFTCFFTCFFTCFFTCFFTCFFTCFFTCFFTCFFTCFFTCFFTCFFT'
      write(*,*)'     Spectral Analysis of Complex Time Series Data '
      write(*,*)'                version 1.02, July 1988'
      write(*,*)'                  (c) ikayex SOFTWARE TOOLS'
      write(*,*)'CFFTCFFTCFFTCFFTCFFTCFFTCFFTCFFTCFFTCFFTCFFTCFFTCFFT'
      write (*,'(///)')
c
c     open files ready for work
            ilog=67
            open( ilog , file='cfft.log')
            rewind(ilog )
            call getdat(iyr,imon,iday)
            call gettim(ihr,imin,isec,i100)
            write(ilog,4)imon,iday,iyr,ihr,imin
 4        format(' @@ DATE:',1x,i2,1h-,i2,1h-,i4,6x,'TIME:',1x,i2,1h:,i2)
c
 40   continue
      write(*,*) ' '
      write(*,*) '  MAIN menu:'
      write(*,*) '          0:   Quit '
      write(*,*) '          1:   PRE-process input data'
      write(*,*) '          2:   FFT transforms'
      write(*,*) '          3:   POST-process transforms'
      write(*,*) '         10:   DOS services'
        write(*,'(a\)')' select -->'
        iopt=intget(0,10)
        write(ilog,*) iopt,' ::MAIN'
        if (iopt.eq.0) stop'stopped from menu'
c
      if (iopt .eq. 1) then
c         dv# is the number of the first element to be associated
          write(ilog,*)'@@  memory: pre-processor'
          dv0 = 1
          dv1 = dv0 + nmax*2
          dv2 = dv1 + nmax*3
          call space(size,dv0,dv2,ierr,ichk)
          write(ilog,*)'@@  used',dv2,' of available',size
          call prep(x,wk,tt,nmax,ilog)
      endif
c
      if (iopt .eq. 2) then
c         dv# is the number of the first element to be associated
          write(ilog,*)'@@  memory: fft transform'
          dv0 = 1
          dv1 = dv0 + nmax*4
          dv2 = dv1 + nmax
          call space(size,dv0,dv2,ierr,ichk)
          write(ilog,*)'@@  used',dv2,' of available',size
          call cfft(x,wk,tt,nmax,ilog)
      endif
c
      if (iopt .eq. 3) then
c         dv# is the number of the first element to be associated
          write(ilog,*)'@@  memory: post-processor'
          dv0 = 1
          dv1 = dv0 + nmax*2
          dv2 = dv1 + nmax*3
          call space(size,dv0,dv2,ierr,ichk)
```

```fortran
            write(ilog,*)'@@  used',dv2,' of available',size
            call post(x,wk,tt,nmax,ilog)
        endif
c
        if (iopt.eq.10) then
            pause' type DOS command or return'
            pause' type DOS command or return'
        endif
c
        go to 40
c
            end
c
            subroutine prep(x,wk,tt,nmax,ilog)
            dimension  x(nmax,2), wk(nmax,2),tt(nmax)
c
   40   continue
        write(*,*)' '
        write(*,*)' PREP menu : '
        write(*,*)'           0: Return'
        write(*,*)'           1: Input time domain data'
        write(*,*)'           2: Interpolate'
        write(*,*)'           3: Signal generator'
        write(*,*)'           4: Separator           '
        write(*,*)'           5: Window data'
        write(*,*)'           6:                      '
        write(*,*)'           7: Exchange columns'
        write(*,*)'           8: Store results'
        write(*,*)'           9: View results '
        write(*,*)'          10: DOS services    '
            write(*,'(a\)')' select -->'
            iopt = intget(0,10)
            write(ilog,*) iopt,' ::PREP'
        if (iopt.eq.0) return
c
        if (iopt.eq.1) then
            call input (wk,tt,ndata,icmp,ilog,nmax)
            icol=icmp+1
c           transfer to 1 and 2 in case already interpolated
            do 50 i=1,ndata
                x(i,1)=wk(i,1)
                x(i,2)=wk(i,2)
   50       continue
            dt=tt(2)-tt(1)
            npt=ndata
          endif
c
        if (iopt.eq.2) then
            write (*,*)' input:  sampling dt | # of lines'
            read (*,*) dt,npt
            write(*,*) dt,npt
            write(ilog,*) dt,npt,'   ::dt, npt'
            tmax  =wk(ndata,3)
            temp = npt*dt
            if (temp.gt.tmax) then
                write (ilog,*) '@@ zeroes added since tmax < npt*dt=',temp
            elseif (temp.lt.tmax) then
                write (ilog,*) '@@ partly used since tmax > npt*dt=',temp
            endif
c
```

```
c          begin interpolation
           call interp(x,wk,tt,      dt,npt,ndata,nmax)
        endif
c
        if (iopt.eq.3) call siggen (x,npt,ilog,dt,nmax)
c
c       average the data
        if (iopt.eq.4) then
c           transfer to 1 and 2 to work array
            do 44 i=1,npt
               wk(i,1)=x(i,1)
               wk(i,2)=x(i,2)
 44         continue
            call averag(wk,tt,npt,icol,ilog,nmax)
        endif
c
c       window the data
        if (iopt.eq.5) then
c           transfer to 1 and 2 to work array
            do 55 i=1,npt
               wk(i,1)=x(i,1)
               wk(i,2)=x(i,2)
 55         continue
            call window (wk,npt,ilog,dt,nmax)
        endif
c
      if (iopt.eq.7) then
c         transfer from working to original
          do 70 i=1,npt
             temp1   = x(i,1)
             temp2   = x(i,2)
             x(i,1) = wk(i,1)
             x(i,2) = wk(i,2)
             wk(i,1)= temp1
             wk(i,2)= temp2
 70       continue
        endif
c
        if (iopt.eq.8) call store (x,wk,dt,df,ilog,nmax,1)
        if (iopt.eq.9) call viewer(x,wk,dt,df,ilog,nmax,1)
        if (iopt.eq.10) then
           pause' Type DOS command or return'
           pause' Type DOS command or return'
        endif
        go to 40
c
          end
c
        subroutine cfft(x,wk,wk1,nmax,ilog)
          dimension x(nmax,2),wk(nmax, 2), wk1(nmax)
          nmax2=nmax/2
c
 40       continue
          write(*,*)' '
          write(*,*)' CFFT menu : '
          write(*,*)'          0: Return'
          write(*,*)'          1: Input time domain data'
          write(*,*)'          2: Forward FFT'
          write(*,*)'          3: Inverse FFT'
          write(*,*)'          4: Input freq domain data'
```

```
        write(*,*)'               5: '
        write(*,*)'               6: '
        write(*,*)'               7: '
        write(*,*)'               8: Store results'
        write(*,*)'               9: View results'
        write(*,*)'              10: DOS services '
        write(*,'(a\)')' select -->'
        read (*,*) iopt
        write(*,*) iopt
        write(ilog,*) iopt, ' ::CFFT'
c
        if (iopt.eq.0) return
        if (iopt.eq.1) then
            call input (wk,wk1,np,icmp,ilog,nmax)
            dt=wk1(2)-wk1(1)
            npt=np
            do 132 i=1,npt
               x(i,1)=wk(i,1)
               x(i,2)=wk(i,2)
132         continue
        endif
c
        if (iopt.eq.2) then
c
        write(*,*)'input:  sampling dt | # of fft points (...8,16,32..)'
        read (*,*) dt,npt
        write(*,*) dt,npt
        write(ilog,*) dt,npt,'  ::dt  npt'
c
        tmax  = wk1(np)
        temp = npt*dt
        if (temp.gt.tmax) then
            write (ilog,*) '@@ zeroes added since tmax < npt*dt=',temp
        elseif (temp.lt.tmax) then
            write (ilog,*) '@@ partly used since tmax > npt*dt=',temp
        endif
c           begin interpolation
            call interp(x,wk   ,wk1,dt,npt,np,nmax)
c           assign to work vector
            do 232 i=1,npt
               wk(i,1)=x(i,1)
               wk(i,2)=x(i,2)
232         continue
            write(ilog,*)'@@ col type ',icmp
            call fftprp (wk(1,1),wk(1,2),wk1(1),wk1(nmax2),
     >                   npt,nfft,2,icmp,ilog,nmax,nmax2)
c           assign to global matrix
            scale=dt
            df=1.0/(nfft*dt)
            if (icmp .eq. 1) df=df*2
            write(ilog,*)'@@  scale df   ',scale,df
            do 270 n=1,npt
               wk(n,1) = wk(n,1)*scale
               wk(n,2) = wk(n,2)*scale
270         continue
        endif
c
        if (iopt.eq.3) then
c           assign to work vector
            do 350 i = 1,npt
```

```
                   x(i,1) = wk(i,1)
                   x(i,2) = wk(i,2)
  350           continue
                write(ilog,*)'@@ col type ',icmp
                call fftprp (x(1,1),x(1,2),wk1(1),wk1(nmax2),
      >                      npt,nfft,3,icmp,ilog,nmax,nmax2)
  c             assign to global matrix
                scale=1.0/(nfft*dt)
                df=scale
                write(ilog,*)'@@  scale dt   ',scale,dt
                do 370 n=1,npt
                   x(n,1) =  x(n,1)*scale
                   x(n,2) =  x(n,2)*scale
  370           continue
  c
           endif
        if (iopt.eq.4) then
            call inreim (wk,wk1,npt,icmp,ilog,dt,nmax)
            df=wk1(2)-wk1(1)
        endif
        if (iopt.eq.8) call store (x,wk,dt,df,ilog,nmax,2)
        if (iopt.eq.9) call viewer(x,wk,dt,df,ilog,nmax,2)
        if (iopt.eq.10) then
            pause'Type DOS command or return'
            pause'Type DOS command or return'
        endif
  c
        go to 40
        end
  c
        subroutine post (x,wk,ff,nmax,ilog)
        dimension  x(nmax,2), wk(nmax,2),ff(nmax)
  c
  40    continue
        write(*,*)' '
        write(*,*)' POST menu : '
        write(*,*)'          0: Return'
        write(*,*)'          1: Input chan AA'
        write(*,*)'          2: Input chan BB'
        write(*,*)'          3: '
        write(*,*)'          4: Bandpass filter input data  '
        write(*,*)'          5: Window input data'
        write(*,*)'          6:                      '
        write(*,*)'          7: Exchange'
        write(*,*)'          8: Store results'
        write(*,*)'          9: View results '
        write(*,*)'         10: Dos services   '
  c     write(*,*)' '
            write(*,'(a\)')' select -->'
            read (*,*) iopt
            write(*,*) iopt
            write(ilog,*) iopt,' ::POST'
        if (iopt.eq.0) return
  c
        if (iopt.eq.1) then
            icol=3
            call inreim(wk,ff,npt,icmp,ilog,dt,nmax)
  c         transfer to 1 and 2 in case already interpolated
            do 50 i=1,npt
               x(i,1)=wk(i,1)
```

```
                  x(i,2)=wk(i,2)
 50            continue
               df=ff(2)-ff(1)
               dfaa=df
c
         elseif (iopt.eq.2) then
               icol=3
               call inreim(wk,ff,npt2,icmp,ilog,dt,nmax)
c              no need to transfer to 1 and 2
               df=ff(2)-ff(1)
               dfbb=df
               df0=abs(dfaa-dfbb)
               if (df0.gt.dfbb/10.0) then
                   write(*,*)'frequency bases of AA BB seem different'
                   write(ilog,*)'@@ frequency bases of AA BB',
     >                          ' seem different',dfaa,dfbb
               endif
         endif
c
      if (iopt.eq.4) then
          write(*,*)'Choose: 1=AA  2=BB'
          ichan=intget(1,2)
          write(ilog,*) ichan,' ::channel'
          if (ichan.eq.1) then
              call filter (x,npt,ilog,nmax)
          else
              call filter (wk,npt,ilog,nmax)
          endif
      endif
c
      if (iopt.eq.5) then
          write(*,*)'Choose: 1=AA  2=BB'
          ichan=intget(1,2)
          write(ilog,*) ichan,' ::channel'
          if (ichan.eq.1) then
              call window (x,npt,ilog,df,nmax)
          else
              call window (wk,npt,ilog,df,nmax)
          endif
      endif
c
      if (iopt.eq.7) then
c         exchange AA and BB
          do 70 i=1,npt
              temp1  =wk(i,1)
              temp2  =wk(i,2)
              wk(i,1)=x(i,1)
              wk(i,2)=x(i,2)
              x(i,1)=temp1
              x(i,2)=temp2
 70       continue
      endif
c
      if (iopt.eq.8) call store (x,wk,dt,df,ilog,nmax,3)
      if (iopt.eq.9) call viewer(x,wk,dt,df,ilog,nmax,3)
      if (iopt.eq.10) then
          pause' type DOS command or return'
          pause' type DOS command or return'
      endif
      go to 40
```

```
c
      end


FFT GROUP

c
      subroutine fftprp (xre,xim,angc,angs,npt,nfft,icom,icmp,ilog,
     >                    nmax,nmax2)
         dimension xre(nmax),xim(nmax), angc(nmax2),angs(nmax2)
         integer*2 ip(4100)
         pi=4*atan(1.0)
c
c        begin fft proper
         call timer('--> FFTprp')
c        prepare for complex form of transform algorithm
         if (icom.eq.2) then
            if (icmp.eq.1) then
c               do if real input
                nfft=npt/2
                do 32 i=1,nfft
                   i2=2*i
                   xre(i)=xre(i2-1)
                   xim(i)=xre(i2)
 32             continue
            else
                nfft=npt
            endif
c
c           form angles table
            write(ilog,*)'@@  nfft   ',nfft
            call timer('--> tables')
            call tables(nfft,angc,angs,ip,nmax,nmax2)
            call timer('--> CMPFFT ')
            call cmpfft(xre,xim,angc,angs,nfft,ip,nmax2)
            call timer('<-- CMPFFT ')
c
            if (icmp.eq.1) then
               arg=pi/float(nfft)
               dcs=cos(arg)
               dsn=sin(arg)
               nfft2=nfft/2
               xre(nfft+1)=xre(1)
               xim(nfft+1)=xim(1)
               ihalf=1
               do 50 i=1,nfft+1
                  nfmi=nfft+2-i
                  iang=(i+1)/2
                  tempr=xre(i)-xre(nfmi)
                  tempi=xim(i)+xim(nfmi)
                  if (ihalf.gt.0) then
                     ang1=angc(iang)
                     ang2=angs(iang)
                     ihalf=-1
                  elseif (ihalf.lt.0) then
                     ang1=angc(iang)*dcs-angs(iang)*dsn
                     ang2=angc(iang)*dsn+angs(iang)*dcs
                     ihalf=1
                  endif
                  xr=(xre(i)+xre(nfmi)) + ang1*tempi-ang2*tempr
```

```
                    xi=(xim(i)-xim(nfmi)) - ang2*tempi-ang1*tempr
                    xr=xr/2
                    xi=xi/2
                    iconj=2*nfft+2-i
                    xre(iconj )=xr
                    xim(iconj )=-xi
50             continue
               do 52 i=1,nfft+1
                    iconj=2*nfft+2-i
                    xre(i)= xre(iconj )
                    xim(i)=-xim(iconj )
52             continue
             endif
          endif
c
          if (icom.eq.3) then
             if (icmp.eq.1) then
c                  do if conjug input
                   nfft=npt/2
                   do 42 i=1,nfft
                      i2=2*i
                      xre(i)=xre(i2-1)+xim(i2-1)
                      xim(i)=xre(i2)   +xim(i2  )
42                 continue
             else
                nfft=npt
             endif
c                  do in both cases
                   do 43 i=1,nfft
                         xim(i)=-xim(i)
43                 continue
c
c          form angles table
           write(ilog,*)'@@  nfft ',nfft
           call timer('--> tables')
           call tables(nfft,angc,angs,ip,nmax,nmax2)
           call timer('--> CMPFFT  ')
           call cmpfft(xre,xim,angc,angs,nfft,ip,nmax2)
           call timer('<-- CMPFFT  ')
c
           do 40 i=1,nfft
                xim(i)=-xim(i)
40         continue
          if (icmp.eq.1) then
            arg=pi/float(nfft)
            dcs=cos(arg)
            dsn=sin(arg)
            nfft2=nfft/2
            xre(nfft+1)=xre(1)
            xim(nfft+1)=xim(1)
            ihalf=1
            do 60 i=1,nfft+1
               nfmi=nfft+2-i
               iang=(i+1)/2
               tempr=xre(i)-xre(nfmi)
               tempi=xim(i)+xim(nfmi)
               if (ihalf.gt.0) then
                  ang1=angc(iang)
                  ang2=angs(iang)
                  ihalf=-1
```

```
                     elseif (ihalf.lt.0) then
                        ang1=angc(iang)*dcs-angs(iang)*dsn
                        ang2=angc(iang)*dsn+angs(iang)*dcs
                        ihalf=1
                     endif
                        if (icom.eq.3) ang2=-ang2
                     xr=(xre(i)+xre(nfmi)) + ang1*tempi-ang2*tempr
                     xi=(xim(i)-xim(nfmi)) - ang2*tempi-ang1*tempr
                     xr=xr/2
                     xi=xi/2
                     iconj=2*nfft+2-i
                     xre(iconj )=xr
                     xim(iconj )=-xi
  60              continue
                  do 62 i=1,nfft+1
                     iconj=2*nfft+2-i
                     xre(i)= xre(iconj )
                     xim(i)=-xim(iconj )
  62              continue
c                 now extract even and odd parts
                  do 64 i=1,npt
                     xre(i)=0.5*(xre(i)-xim(i))
                     xim(i)=0.0
  64              continue
                endif
c
              endif
              call timer('<-- FFTprp')
c
              return
              end
c
c based on  Brighams FFT pp.164
              subroutine cmpfft(xreal,ximag,angc,angs,npt,ip,nmax2)
                 dimension xreal(npt),ximag(npt),angc(nmax2),angs(nmax2)
                 integer*2 ip(npt)
                 integer p
c
                 pi=4*atan(1.0)
c                icntp=# of loops,     icnts=# of complex mults
                 icntp=0
                 icnts=0
c
                 nu=ifix(alog(float(npt))/alog(float(2))+0.5)
                 n22=npt/4
c
c          fft do loops
              write(*,*)' '
              k=0
              num1=nu
              n2=npt
              do 100 level=1,nu
                  write(*,'(''+ level = '',i2)') level
                  k=0
                  num1=num1-1
                  n2=n2/2
  102           continue
                do 101 i=1,n2
                    k1=k+1
                    k1n2=k1+n2
```

```
                kp=k/n2
                p=ip(kp+1)
                icntp=icntp+1
c               do the complex multiplications
                if (p.eq.0) then
                    treal=xreal(k1n2)
                    timag=ximag(k1n2)
                elseif (p.eq.n22) then
                    treal= ximag(k1n2)
                    timag=-xreal(k1n2)
                else
                    icnts=icnts+1
                    cn=angc(p+1)
                    sn=angs(p+1)
                    treal=xreal(k1n2)*cn+ximag(k1n2)*sn
                    timag=ximag(k1n2)*cn-xreal(k1n2)*sn
                endif
c
                xreal(k1n2)=xreal(k1)-treal
                ximag(k1n2)=ximag(k1)-timag
                xreal(k1)=xreal(k1)+treal
                ximag(k1)=ximag(k1)+timag
c
                k=k+1
  101           continue
c
            k=k+n2
            if(k.lt.npt) goto 102
c
  100       continue
c
c           unscramble the transform
            do 103   k=1,npt
                i=ip(k)+1
                if (i.le.k) goto 103
                treal=xreal(k)
                timag=ximag(k)
                xreal(k)=xreal(i)
                ximag(k)=ximag(i)
                xreal(i)=treal
                ximag(i)=timag
  103       continue
c
            write(*,*)'counts: loops  mults',icntp,icnts
            return
            end
c
            subroutine tables(nfft,angc,angs,ip,nmax,nmax2)
                dimension angc(nmax2),angs(nmax2)
                integer*2 ip(nmax)
                pi=4.0*atan(1.0)
c
                n2=nfft/2
                n22=nfft/4
c
c               angles
                arg=2*pi/float(nfft)
                angc(n22+1)=0.0
                angs(n22+n22+1)=angc(n22+1)
                angc(1)=1.0
```

```
             angs(1)=0.0
             angc(n2+1)=-1.0
             angs(n22+1)=1.0
             argp=0.0
          do 29 i=2,n22
             argp=argp+arg
             cssn=cos(argp)
             angc(i)=cssn
             angc(n2-i+2)=-cssn
             angs(n22+i)= cssn
             angs(n22-i+2)= cssn
29        continue
c         bit values
          nu=ifix(alog(float(nfft))/alog(float(2))+0.5)
          ip(1)=0
          nd2=nfft
          do 31 nnu=0,nu-1
             nnu2=2**nnu
             nd2=nd2/2
             do 39 i=1,nnu2
                ip(nnu2+i)=ip(i)+nd2
39           continue
31        continue
c
          return
          end
```

I/O GROUP

```
c
       subroutine input (wk,tt,np,icmp,ilog,nmax)
          dimension wk(nmax, 2),tt(nmax)
          character*30 infile
c
          write(*,'(a\)')' type filename -->'
          read(*,'(1a30)') infile
          write(*,*) infile
          write(ilog,*) infile,'     ::filename'
          write(*,*)'Input: # of lines | 1=re 2=re-pad 3=cmplx'
     >                            ,' | 1=asc 2=unf 3=bin'
          read (*,*) np,icol,ibin
          write(*,*) np, icol,ibin
          write(ilog,*) np, icol,ibin,'  ::npt icol bin'
          if (icol.eq.1) icmp=1
          if (icol.gt.1) icmp=2
c
          if (ibin.eq.1) then
             open (9,file=infile,iostat=ierr,err=170)
          elseif (ibin.eq.2) then
             open (9,file=infile,form='unformatted',
     >                                  iostat=ierr,err=170)
          elseif (ibin.eq.3) then
             open (9,file=infile,form='binary',
     >                                  iostat=ierr,err=170)
          endif
          rewind 9
c
          if (ibin.eq.1) then
             write(*,*)' reading <ascii> file'
```

```
          do 60 i = 1,np
             if (icol.eq.1.or.icol.eq.2) then
                 read (9,*,end=160) tt(i) ,wk(i,1)
                 wk(i,2)=0.0
             else
                 read (9,*,end=160) tt(i) ,wk(i,1),wk(i,2)
             endif
60        continue
       else
          write(*,*)' reading <unf or bin> file'
          do 62 i = 1,np
             if (icol.eq.1.or.icol.eq.2) then
                 read (9,end=160) tt(i) ,wk(i,1)
                 wk(i,2)=0.0
             else
                 read (9,end=160) tt(i) ,wk(i,1),wk(i,2)
             endif
62        continue
       endif
       close(9)
c
c      adjust times
       tzero = tt(1)
       do 10 i = 1,np
          tt(i)  = tt(i) -tzero
10     continue
c
       return
c
c      ERRORS
160    write(*,*)' !!! fewer data lines than expected'
       np=i-1
       write(*,*)'# of data =',np
       write(ilog,*)'@@  # of data =',np
       close(9)
       return
170    if (ierr.eq.2) then
          write (6,180) infile
180       format (' ...can not find file: ',1a30,/)
       else
          write (6,190) eropen,infile
190       format (' ...error ',i3,' occurred in opening file: ',1a30)
       endif
c
200    stop'stopped due to read error'
       end
c
     subroutine inreim (wk,ff,npt,icmp,ilog,dt,nmax)
       dimension wk(nmax,2),ff(nmax)
       character*30 infile
c
       write(*,'(a\)')' type filename (freq re im) -->'
       read(*,'(1a30)') infile
       write(*,*) infile
       write(ilog,*) infile,'    ::filename'
       write(*,*)'input:  # of fft lines | 1=cnj 2=pad 3=cmplx '
     >                      ,'| 1=asc 2=unf 3=bin'
       read (*,*) npt,icol,ibin
       write(*,*) npt, icol,ibin
       write(ilog,*) npt, icol,ibin,'  ::npt icl bin '
```

```
          if (icol.eq.1) icmp=1
          if (icol.gt.1) icmp=2
          ndata=npt
          nyqst=npt/2+1
          if (icol.le.2) ndata=nyqst
c
          if (ibin.eq.1) then
              open (8,file=infile,iostat=ierr,err=70)
          elseif (ibin.eq.2) then
              open (8,file=infile,form='unformatted',
     >                                      iostat=ierr,err=70)
          elseif (ibin.eq.3) then
              open (8,file=infile,form='binary',
     >                                      iostat=ierr,err=70)
          endif
          rewind 8
c
          if (ibin.eq.1) then
              do 50 i = 1,ndata
                  read (8,*,end=96) xjunk ,wk(i,1),wk(i,2)
                  ff(i)=xjunk
 50           continue
          else
              do 52 i = 1,ndata
                  read (8,end=96) xjunk ,wk(i,1),wk(i,2)
                  ff(i)=xjunk
 52           continue
          endif
          close(8)
c
c         complete conjugate arrays in case
          if (icol.le.2) then
              do 60 n=1,nyqst-1
                  wk(npt+2-n,1)= wk(n,1)
                  wk(npt+2-n,2)=-wk(n,2)
 60           continue
          endif
          dt = float(ndata-1)/float(npt )/xjunk
          write(*,*)'dt from reim',dt
         return
c
c         ERRORS
 96       write(*,*)' !!! fewer lines  than expected'
            npt=i-1
            write(*,*)'# of data =',npt
            dt = float(npt-1)/float(npt)/xjunk
            write(*,*)'dt from reim',dt
            write(ilog,*)'@@  dt from reim error',dt
            write(ilog,*)'@@  # of data =',npt
          return
 70       if (ierr.eq.2) then
              write (6,80) infile
 80           format (' ...can not find file: ',1a30,/)
          else
              write (6,90) eropen,infile
 90           format (' ...error ',i3,' occurred in opening file: ',1a30)
          endif
c
          stop'stopped due to read error '
          end
```

```
c
      subroutine store (x,wk,dt,df,ilog,nmax,imen)
      dimension x(nmax,2),wk(nmax,2), y(21)
      character*30 outfil
c
10    write(*,*)' '
      write(*,*)'STORE'
      write(*,*)'Choose:  0=return, 1=store asc, 2=unf, 3=bin'
      read (*,*) istore
      write(ilog,*) istore,' ::store'
      if (istore.eq.0) return
         if (imen.eq.1) then
            i12=15
            i34=15
            i56=15
            i78=15
         elseif (imen.eq.2) then
            i12=15
            i34=16
            i56=15
            i78=15
         elseif (imen.eq.3) then
            i12=16
            i34=16
            i56=16
            i78=16
         endif
c
      write(*,'(a\)')' type outfilename -->'
      read(*,'(1a30)') outfil
      write(*,*) outfil
      write(ilog,*) outfil,'  ::filename'
      if (istore.eq.1) then
         open (7,file=outfil,iostat=ierr,err=100)
      elseif (istore.eq.2) then
         open (7,file=outfil,form='unformatted',iostat=ierr,err=100)
      elseif (istore.eq.3) then
         open (7,file=outfil,form='binary',iostat=ierr,err=100)
      endif
      rewind 7
c
      write(*,*)' STORE menu : '
      call menu(imen)
      write(*,*)' input:  col1 | col2 | # of data | interval'
      write(*,*)'         (combinations 1-2,3-4,5-6,7-8 have t or f) '
      read (*,*) is1,is2,n,nint
      write(*,*) is1,is2,n,nint
      write(ilog,*) is1,is2,n,nint,' ::col1,col2, n, int'
c
         if (istore.eq.1) then
            write(*,*)'writing <ascii> file'
         else
            write(*,*)'writing <unf or bin> file'
         endif
         do 40 i = 1,n,nint
c        fill arrays
            y(1) = x(i,1)
            y(2) = x(i,2)
            y(3) = wk(i,1)
            y(4) = wk(i,2)
```

```
                  call array(y,imen)
                  y(15) = real(i-1)*dt
                  y(16) = real(i-1)*df
            if (istore.eq.1) then
                if (is1.eq.1.and.is2.eq.2) then
                    write (7,*) y(i12  ),y(is1),y(is2)
                elseif (is1.eq.3.and.is2.eq.4) then
                    write (7,*) y(i34  ),y(is1),y(is2)
                elseif (is1.eq.5.and.is2.eq.6) then
                    write (7,*) y(i56  ),y(is1),y(is2)
                elseif (is1.eq.7.and.is2.eq.8) then
                    write (7,*) y(i78  ),y(is1),y(is2)
                else
                    write (7,*) y(is1),y(is2)
                endif
            else
                if (is1.eq.1.and.is2.eq.2) then
                    write (7) y(i12  ),y(is1),y(is2)
                elseif (is1.eq.3.and.is2.eq.4) then
                    write (7) y(i34  ),y(is1),y(is2)
                elseif (is1.eq.5.and.is2.eq.6) then
                    write (7) y(i56  ),y(is1),y(is2)
                elseif (is1.eq.7.and.is2.eq.8) then
                    write (7) y(i78  ),y(is1),y(is2)
                else
                    write (7) y(is1),y(is2)
                endif
            endif
 40         continue
            close(7)
            goto 10
c
        return
c
c       ERRORS
 100    write (*,*) ' ...error occurred in opening file: ',outfil
        go to 10
c
      end
c
      subroutine viewer (x,wk,dt,df,ilog,nmax,imen)
        dimension x(nmax,2),wk(nmax,2), y(21)
c
 10     write(*,*)' '
        write(*,*)'VIEWER'
        write(*,*)'Choose:  0=return, 1=print, 2=plot'
        iprint=intget(0,2)
        write(ilog,*) iprint,' ::print'
        if (iprint.eq.0) then
            return
        elseif (iprint.eq.2) then
            open (17,file='scrndata.bin',form='binary',err=100)
            rewind 17
        endif
c
        write(*,*)' '
        write(*,*)'VIEWER menu : '
        call menu(imen)
        write(*,*)' '
        write(*,*)'input:  col1 | col2 | # of data | interval'
```

```
        write(*,*)' '
        read (*,*) is1,is2,n,nint
        write(*,*) is1,is2,n,nint,' col1 | col2 | # of pt | inter'
        write(*,*)' '
        write(ilog,*) is1,is2,n,nint,' ::col1,col2, # of pt, inter'
c
        do 40 i = 1,n,nint
c           fill arrays
            y(1) = x(i,1)
            y(2) = x(i,2)
            y(3) = wk(i,1)
            y(4) = wk(i,2)
            call array(y,imen)
            y(15) = real(i-1)*dt
            y(16) = real(i-1)*df
            if (iprint.eq.1) then
                write (*,*) i,y(is1),y(is2)
            elseif (iprint.eq.2) then
                write (17) y(is1),y(is2)
            endif
40      continue
        if (iprint.eq.2) then
            close (17)
            call screen
        endif
        goto 10
c
100     write (*,*) ' ...error occurred in opening screen file '
        go to 10
        end
c
        subroutine menu(imen)
        if (imen.eq.1) then
c           sep
            write(*,*)'        1=real in           2=imag in         '
            write(*,*)'        3=real work         4=imag work        '
            write(*,*)'        5=real diff         6=imag diff        '
            write(*,*)'  '
            write(*,*)'        9=abs in           10=abs work         '
            write(*,*)'       15=time             16=freq   '
        elseif (imen.eq.2) then
c           fft
            write(*,*)'        1=real TDD          2=imag TDD  '
            write(*,*)'        3=real FDD          4=imag FDD  '
            write(*,*)'  '
            write(*,*)'        9=amp FDD          10=amp FDD (log)'
            write(*,*)'       11=phase '
            write(*,*)'       15=time             16=freq    '
            write(*,*)' '
        elseif (imen.eq.3) then
c           post
            write(*,*)'        1=real AA           2=imag AA         '
            write(*,*)'        3=real BB           4=imag BB         '
            write(*,*)'        5=real x-PSD        6=imag x-PSD      '
            write(*,*)'        7=real CONV         8=imag CONV  '
            write(*,*)'        9=real DECONV      10=imag DECONV  '
            write(*,*)'  '
            write(*,*)'       11=PSD  AA          12=PSD (log10)    '
            write(*,*)'       13=PSD  BB          14=PSD (log10)    '
            write(*,*)'       15=time             16=freq     '
```

```
                  write(*,*)' '
               endif
c
               return
               end
c
            subroutine array(y,imen)
               dimension y(21)
               complex aaa,bbb,ccc,ddd
c
               if (imen.eq.1) then
                   y(5) = y(1) - y(3)
                   y(6) = y(2) - y(4)
                   y(9) = sqrt(y(1)*y(1)+y(2)*y(2))
                  y(10) = sqrt(y(3)*y(3)+y(4)*y(4))
               elseif (imen.eq.2) then
                   x5 = y(3)
                   x6 = y(4)
                   dumam=sqrt(x5*x5+x6*x6)
                   dvalue=dumam
                   y(9) = dvalue
                   y(10) = alog10(dvalue+1e-12)
                   if (x5.eq.0.0.and.x6.eq.0.0) then
                       theta = 0.0
                   else
                       theta=atan2(x6,x5)
                   endif
                   y(11) = theta
               elseif (imen.eq.3) then
                   y( 5) = y(1)*y(3) + y(2)*y(4)
                   y( 6) =-y(1)*y(4) + y(2)*y(3)
                     aaa = y(1)+ci*y(2)
                     bbb = y(3)+ci*y(4)
                     if (cabs(aaa).le.1e-12) then
                          aaa=1e-12
                     endif
                     ccc = bbb*aaa
                     ddd = bbb/aaa
                   y( 7) = real(ccc)
                   y( 8) = aimag(ccc)
                   y( 9) = real(ddd)
                   y(10) = aimag(ddd)
                   y(11) = y(1)*y(1) + y(2)*y(2)
                   y(12) = alog10(y(11)+1e-12)
                   y(13) = y(3)*y(3) + y(4)*y(4)
                   y(14) = alog10(y(13)+1e-12)
               endif
c
               return
               end
```

PREP GROUP

```
c
            subroutine averag(y,tt,npt,icol,ilog,nmax)
               dimension  y(nmax,2),tt(nmax)
c
 522        continue
            write(*,*)'SEPARATE'
```

```
              write(*,*)' choose: 0=return, 1=separate'
              iaver=intget(0,1)
              write(ilog,*) iaver,' :: averaging'
              if (iaver.eq.0) return
c
              write(*,*)'input: # of averages | starting point'
              read (*,*) navg,nutd
              write(*,*) navg,nutd
              write(ilog,*) navg,nutd,' ::# of avg, starting pt.'
c
              n0=(navg-1)/2
c
c             1=orig(or prev)  3=mean
              do 590 j=1,icol-1
                 do 505 n=1,nutd
                    tt(n) =y(n,j)
 505             continue
c
                 do 510 n=nutd+1,nutd+n0
                    sum=0.0
                    n2m1=(n-nutd)*2-1
                    do 512 ns=nutd+1,nutd+n2m1
                       sum=sum+y(ns,j)
 512                continue
                    tt(n) =sum/n2m1
 510             continue
c
                 do 530 n=nutd+n0+1,npt+n0
                    sum=0.0
                    n1=n-n0
                    n2=n+n0
                    do 532 ns=n1,n2
                       sum=sum+y(ns,j)
 532                continue
c more efficient sum = sum -y(n1,1) + y(n2,1)
                    tt(n) =sum/navg
 530             continue
c
c                update arrays
                 do 550 n=1,npt+n0
                    y(n,j)=tt(n)
 550             continue
c
 590          continue
              goto 522
              return
              end
c
          subroutine interp(x,wk,tt,dt,npt,np,nmax)
            dimension x(nmax,2),wk(nmax,2),tt(nmax)
c
            x(1,1) = wk(1,1)
            x(1,2) = wk(1,2)
            tmax   = tt(np)
            time = 0.0
            k = 1
            j = 1
 80         time = time+dt
            if (k.eq.npt.or.time.gt.tmax) goto 120
              k = k+1
```

```
              do 110 i = j,np
                if (tt(i).lt.time) go to 110
                if (i.eq.1) then
      90            x(k,1) = wk(1,1)/tt(1)*time
                    x(k,2) = wk(1,2)/tt(1)*time
                else
                    dtdt =   (time-tt(i-1))/(tt(i)-tt(i-1))
                    x(k,1) = (wk(i,1)-wk(i-1,1))*dtdt + wk(i-1,1)
                    x(k,2) = (wk(i,2)-wk(i-1,2))*dtdt + wk(i-1,2)
                endif
                j = i
                go to 80
  110         continue
c
 120          continue
c             pad remainder with zeroes
              do 130 i = (k+1),npt
                x(i,1) = 0.0
                x(i,2) = 0.0
 130          continue
c
              return
              end
c
        subroutine window(wk,npt,ilog,dt,nmax)
          dimension wk(nmax,2)
          pi=4*atan(1.0)
c
          write(*,*)'WINDOW menu:'
          write(*,*)'              1: Triangular'
          write(*,*)'              2: Trapazoid '
          write(*,*)'              3: Exponential'
          write(*,*)'              4: Hanning   '
          write(*,*)'              5: Gaussian'
          write(*,*)'              6: '
          write(*,'(a\)')' select -->'
          iwin=intget(1,5)
          write(*,*) iwin
          write(ilog,*) iwin,' ::window type'
        write(*,*)' input:  constants  1 |  2 |  3 |  4 '
        read(*,*) tt1,tt2,tt3,tt4
        write(*,*) tt1,tt2,tt3,tt4
        write(ilog,*) tt1,tt2,tt3,tt4,'  ::const'
c
        do 50 n=1,npt
          ttt=real(n-1)*dt
          if (iwin.eq.1) then
            if (ttt.lt.tt1 ) then
              scale = 0.0
            else if (ttt.ge.tt1 .and. ttt.lt.tt2) then
              scale = (ttt-tt1)/(tt2-tt1)
            else if (ttt.ge.tt2 .and. ttt.lt.tt3) then
              scale = (tt3-ttt)/(tt3-tt2)
            else if (ttt.ge.tt3 .and. ttt.lt.tt4) then
              scale = 0.0
            else if (ttt.ge.tt4) then
              scale = 0.0
            endif
          elseif (iwin.eq.2) then
            if (ttt.lt.tt1 ) then
```

```
              scale = 0.0
          else if (ttt.ge.tt1 .and. ttt.lt.tt2) then
              scale = (ttt-tt1)/(tt2-tt1)
          else if (ttt.ge.tt2 .and. ttt.lt.tt3) then
              scale = 1.0
          else if (ttt.ge.tt3 .and. ttt.lt.tt4) then
              scale = (tt4-ttt)/(tt4-tt3)
          else if (ttt.ge.tt4) then
              scale = 0.0
          endif
      elseif (iwin.eq.3) then
          if (ttt.lt.tt1 ) then
              scale = 0.0
          else if (ttt.ge.tt1 .and. ttt.lt.tt2) then
              scale = (ttt-tt1)/(tt2-tt1)
          else if (ttt.ge.tt2 .and. ttt.lt.tt3) then
              scale = 1.0
          else if (ttt.ge.tt3 ) then
              scale=exp(-3.*(ttt-tt3)/(tt4-tt3))
          endif
      elseif (iwin.eq.4) then
          if (ttt.lt.tt1 ) then
              scale = 0.0
          else if (ttt.ge.tt1 .and. ttt.lt.tt2) then
              tts = (ttt-tt1)/(tt2-tt1)
              scale = (sin(pi*tts/2))**2
          else if (ttt.ge.tt2 .and. ttt.lt.tt3) then
              scale = 1.0
          else if (ttt.ge.tt3 .and. ttt.lt.tt4) then
              tts = (ttt-tt3)/(tt4-tt3)
              scale = 1.0 - (sin(pi*tts/2))**2
          else if (ttt.ge.tt4) then
              scale = 0.0
          endif
      elseif (iwin.eq.5) then
          if (ttt.lt.tt2 ) then
              tts = (ttt-tt2)/(tt2-tt1)
              scale=exp(-tts*tts)
          else if (ttt.ge.tt2 .and. ttt.lt.tt3) then
              scale = 1.0
          else if (ttt.ge.tt3 ) then
              tts = (ttt-tt3)/(tt4-tt3)
              scale=exp(-tts*tts)
          endif
      endif
      wk(n,1)=wk(n,1)*scale
      wk(n,2)=wk(n,2)*scale
50    continue
c
      return
      end
c
      subroutine siggen(x,npt,ilog,dt,nmax)
      dimension x(nmax,2)
      pi=4*atan(1.0)
c
      write(*,*)' input:  dt | # of pts '
      read(*,*) dt,npt
      write(*,*) dt,npt
      write(ilog,*) dt,npt ,' :: dt   # of pts'
```

```
c
        write(*,*)' signal = sin(c1 + c2*t + c3*t**c4)*2pi'
        write(*,*)' input:  coeffs  c1 |  c2 |  c3 |  c4 '
        read(*,*) c1,c2,c3,c4
        write(*,*) c1,c2,c3,c4
        write(ilog,*) c1,c2,c3,c4,'  ::coeffs '
c
        do 50 n=1,npt
           ttt=real(n-1)/real(npt-1)*pi*2
           tts=(c1 + c2*ttt + c3*ttt**c4)
           signal=sin(tts)
              x(n,1)=signal
              x(n,2)=signal
  50    continue
c
        return
        end
c
        subroutine filter(wk,npt,ilog,nmax)
          dimension wk(nmax, 2)
c
  40    continue
        write(*,*)'FILTER'
        write(*,*)'choose:  0=return  1=filter'
        ifilt=intget(0,1)
        write(ilog,*) ifilt,'  ::filter'
        if (ifilt.eq.0) return
c
        write(*,*)'input: nlower | nhigher  of 000 filtered range'
        read (*,*) nlow,nhigh
        write(ilog,*) nlow,nhigh,'  ::low  high'
        nnyq=npt/2+1
        if (nhigh.le.nnyq) goto 49
           write(*,*)'nhigher is greater than nyquist, will use nyqst'
           write(ilog,*)'@@  nhigher >  nyquist, will use nyqst'
           nhigh=nnyq
  49    continue
        do 50 n=nlow,nhigh
           wk(n,1)=0
           wk(n,2)=0
           wk(npt+2-n,1)=0
           wk(npt+2-n,2)=0
  50    continue
c
        goto 40
        return
        end
```

UTILITY GROUP

```
c
        interface to integer*2 function system [c]
   >        (string[reference])
            character*1 string
        end
c
        subroutine screen
          integer*2 system
          i=system('scrnbat  'c)
```

```
          return
          end
c
          subroutine timer(message)
            integer*2 ihr,imin,isec,i100
            character*10 message
            call gettim(ihr,imin,isec,i100)
                it1 = 3600*ihr + 60*imin + isec
                write(*,10) it1,i100,message
  10            format(1x,i5,1h.,i2,3hsec,2x,1a10)
          return
          end
c
          subroutine counter(i)
            icount=icount+1
            if (icount.eq.20) then
                write(*,'(''+ counter ='',i4)') i
                icount=0
            endif
          return
          end
c
c     This checks the space requirements
      subroutine space(size,start,iend,ierr,ichk)
        integer size, start, iend, ierr
c
      if (size.le.iend) then
         write(err,9001) iend
         ichk=0
         return
      endif
         ichk=1
         write(*,*) 'used',iend, ' of available',size
c
c           ERROR MESSAGES
 9001 format('SPACE ERROR: Not enough space in real matrix ',
     >          'set size in main program to be larger than ', i10)
c
      return
      end
c
c     FUNCTION INTGET
      integer function intget(min, max)
        integer min, max, line
c
      do 10 i= 1, 3
         read(*,*) line
         write(*,*) line
         if (line .lt. min) then
            write(*,*)' **** Minimum possible is ',min,', try again.'
         elseif (line .gt. max) then
            write(*,*)' **** Maximum possible is ',max,', try again.'
         else
            intget = line
            return
         endif
  10  continue
      write(*,*)' Can not get acceptable answer. Assuming minimum:',min
      intget = min
      return
```

```
      end
c
c FUNCTION REALGET
      real function realget(min, max)
        real min, max, line
c
      do 10 i= 1,3
         read(*,*) line
         write(*,*) line
         if (line .lt. min) then
            write(*,*)' **** Minimum possible is ',min,', try again.'
         elseif (line .gt. max) then
            write(*,*)' **** Maximum possible is ',max,', try again.'
         else
            realget = line
            return
         endif
 10      continue
      write(*,*)' Can not get acceptable answer. Assuming minimum:',min
      realget = min
      return
      end
c
```

PROPCOMP

```
c         PROPCOMP.for : bare bones propagation program.
c
          dimension tran(4100,3)
          complex ci,ecmplx,fcmplx,ocmplx,tf1,tf2
          complex xk,betac,eomega,arg1,arg2
          character*30 file1,file3
c
          ci=cmplx(0,1)
          pi=4.0*atan(1.0)
c
          write(*,'(//)')
          write(*,*)'PROPPROPPROPPROPPROPPROPPROPPROPPROPPROPPROP '
          write(*,*)'                  version 2.06, August 1988 '
          write(*,*)'                    ikayex SOFTWARE TOOLS'
          write(*,*)'PROPPROPPROPPROPPROPPROPPROPPROPPROPPROPPROP '
          write(*,'(//)')
c
c         MATERIAL PROPERTIES
              e1=10.6e6
              rho1=2.7e-4
              width1=1
              thick1=0.25
              write(*,*)'e1 etc',e1,rho1,width1,thick1
              area1 = thick1*width1
              zi=1/12.*width1*thick1**3
              ea=e1*area1
              c0=sqrt(e1/rho1)
              freq0=10000.
              omega0=freq0*2*pi
c
c         READ INPUT FILE
          write(*,'(a\)')' type:  input filename--> '
          read(*,'(1a30)') file1
```

```
          write(*,*) file1
          write(*,*)' # lines,  in-pos'n, out-posn  '
          read(*,*) ndata,xx1,xx2
          write(*,*) ndata,xx1,xx2
          nyqst=ndata/2+1
c
              open(unit=3,file=file1)
              write(*,*)'reading <ascii> file'
              rewind 3
              do 20 nn=1,ndata
                 read(3,*) tran(nn,1),tran(nn,2),tran(nn,3)
20            continue
              close (unit=3)
c
          write(*,*)'Types of problems:'
          write(*,*)'                    1=non-dispersive'
          write(*,*)'                    2=dispersive rod'
          write(*,*)'                    3=dispersive beam'
          write(*,*)'                    4=strain to displacement'
          write(*,*)'                    5=strain to velocity'
          write(*,'(a\)')' select one--> '
          read(*,*) itype
          write(*,*)itype
c
c         BIG DO_LOOP OVER FREQUENCIES only up to Nyquist
          do 30 nn=2,nyqst
             freq=tran(nn,1)
             omega = tran(nn,1)*2.0*pi
             eomega=e1
c
c            simple transmission
             if (itype.eq.1) then
                xk=omega*csqrt(rho1/eomega)
                arg1=xk*xx1
                arg2=xk*xx2
                tf1=-cexp(-ci*arg1)
                tf2=-cexp(-ci*arg2)
c
c               obtain output
                fcmplx = tran(nn,2)+ci*tran(nn,3)
                ecmplx = fcmplx*tf2/tf1
                ocmplx = ecmplx
c
c            dispersion rod transmission
             elseif (itype.eq.2) then
                xk=(omega0+omega)/(0.5*omega0+1.0*omega)*omega/c0
                arg1=xk*xx1
                arg2=xk*xx2
                tf1=-cexp(-ci*arg1)
                tf2=-cexp(-ci*arg2)
c
c               obtain output
                fcmplx = tran(nn,2)+ci*tran(nn,3)
                ecmplx = fcmplx*tf2/tf1
                ocmplx = ecmplx
c
c            dispersion  beam transmission
             elseif (itype.eq.3) then
                xk=sqrt(omega*omega0)/c0
                arg1=xk*xx1
```

```
                   arg2=xk*xx2
                   tf1=-cexp(-ci*arg1)
                   tf2=-cexp(-ci*arg2)
c
c                  obtain strain
                   fcmplx = tran(nn,2)+ci*tran(nn,3)
                   ecmplx = fcmplx*tf2/tf1
                   ocmplx = ecmplx
c
c              dispersion beam  strain to displacement
               elseif (itype.eq.4) then
                   xk=sqrt(omega)*(rho1*area1/e1/zi)**0.25
                   arg1=xk*xx1
                   arg2=xk*xx2
                   tf1=-cexp(-ci*arg1)-ci*cexp(-arg1)
                   tf1=tf1*xk*xk
                   tf2= cexp(-ci*arg2)-ci*cexp(-arg2)
c
c                  obtain strain
                   fcmplx = tran(nn,2)+ci*tran(nn,3)
                   ecmplx = fcmplx*tf2/tf1
                   ocmplx = ecmplx*2/thick1
c
c              dispersion beam  strain to velocity
               elseif (itype.eq.5) then
                   xk=sqrt(omega)*(rho1*area1/e1/zi)**0.25
                   arg1=xk*xx1
                   arg2=xk*xx2
                   tf1=-cexp(-ci*arg1)-ci*cexp(-arg1)
                   tf1=tf1*xk*xk
                   tf2= cexp(-ci*arg2)-ci*cexp(-arg2)
                   tf2= tf2*ci*omega
c
c                  obtain velocity from strain
                   fcmplx = tran(nn,2)+ci*tran(nn,3)
                   ecmplx = fcmplx*tf2/tf1
                   ocmplx = ecmplx*2/thick1
c
               endif
c
c              FORM THE COMPLEX CONJUGATE
               tran(nn,2)         = real(ocmplx)
               tran(nn,3)         = aimag(ocmplx)
               tran(2+ndata-nn,2) = tran(nn,2)
               tran(2+ndata-nn,3) =-tran(nn,3)
 30        continue
c
c          ASSIGN ZERO TO INITIAL VALUE
           sum=0.0
           do 32 n=2,nyqst
               sum=sum+tran(n,2)
 32        continue
           tran(1,2)          = -2*sum
           tran(1,3)          = 0.0
           tran(2+ndata-1,2) = tran(1,2)
           tran(2+ndata-1,3) =-tran(1,3)
c
c          STORE RESULTS
 800       continue
           write(*,'(a\)')'type:  output filename--> '
```

```
      read(5,'(1a30)') file3
      write(*,*) file3
         open(unit=2,file=file3)
         write(*,*)'storing <ascii> file'
         rewind 2
         do 820 nn=1,ndata
            write(2,*) tran(nn,1),tran(nn,2),tran(nn,3)
820      continue
         close(2)
c
      stop 'terminated ok'
      end
```

RODCOMP

```
c        RODCOMP.for : wave propagation in rods.
c
         dimension tran(4100,3)
         complex a(4,4),b(4),d(5)
         complex ci,ecmplx,fcmplx,ocmplx,tf1,tf2,tff
         complex eomega,xk,arg
         complex cex2m,cex2p,cex3m
         character*30 file2,file3
c
         ci=cmplx(0,1)
         pi=4*atan(1.0)
c
         write(*,'(//)')
         write(*,*)'RODRODRODRODRODRODRODRODRODRODRODRODRODRODRODROD    '
         write(*,*)'              version 2.05, August 1988'
         write(*,*)'            ikayex SOFTWARE TOOLS  '
         write(*,*)'RODRODRODRODRODRODRODRODRODRODRODRODRODRODRODROD    '
c        open log file
         ilog=67
         open(ilog,file='rod.log')
         rewind(ilog)
c
900      continue
         write(*,*)'   '
         write(*,*)'      MAIN menu:        '
         write(*,*)'           0 :  quit    '
         write(*,*)'           1 :  read material props    '
         write(*,*)'           2 :  read input file        '
         write(*,*)'           4 :  do wave analysis   '
         write(*,*)'           8 :  store results        '
         write(*,'(a\)')' select one--> '
         read(5,*) iopt
         write(ilog,*) iopt,' ::MAIN'
c
c        quit
         if (iopt.eq.0) then
            stop'stopped from menu'
c
c        read rod material props
         elseif (iopt.eq.1) then
            write(*,*)' e1 rho1 znu1 '
            write(*,*)' dim11  dim12  '
            write(*,*)' e2 rho2 znu2 '
            write(*,*)' dim21  dim22  '
```

```
                  write(*,*)' eta1   eta2'
                  write(*,*)' zmass'
                  read(*,*) e1,rho1,znu1
                  read(*,*) width1,thick1
                  read(*,*) e2,rho2,znu2
                  read(*,*) width2,thick2
                  read(*,*) eta1, eta2
                  read(*,*) zmass
                  write(ilog,*) e1,rho1,znu1,' :: e rho nu  1'
                  write(ilog,*) width1,thick1,' :: dim1 dim2   1'
                  write(ilog,*) e2,rho2,znu2,' :: e rho nu  2'
                  write(ilog,*) width2,thick2,' :: dim1 dim2   2'
                  write(ilog,*) eta1, eta2,' :: eta1 eta2    '
                  write(ilog,*) zmass,' :: mass'
c
c        read input file
          elseif (iopt.eq.2) then
                  write(*,*)'READFILE'
                  write(*,*)'Choose:  0=return, 1=read asc, 2=unf, 3=bin'
                  read(*,*) iread
                  write(ilog,*) iread,' :: iread'
                      if (iread.eq.0) goto 900
                      write(*,'(a\)')' type:  input filename--> '
                      read(*,'(1a30)') file3
                      write(ilog,*) file3,' ::filename'
                      write(*,*)'type :    # data,  in-pos'n, fscale  '
                      read(*,*) ndata, xx2, fscale
                      read(*,*) xbnd , blen
                      write(ilog,*) ndata, xx2,fscale,':: ndata x2 fscale'
                      write(ilog,*) xbnd,blen ,':: xbnd,blen'
                  call fylopn(file3,iread)
                  nyqst=ndata/2+1
                  if (iopen.eq.0) then
                      write(ilog,*)'QQ reading <ascii> file'
                      do 20 nn=1,nyqst
                          read(9,*) tran(nn,1),tran(nn,2),tran(nn,3)
  20                  continue
                  else
                      write(ilog,*)'QQ reading <unf or bin> file'
                      do 22 nn=1,nyqst
                          read(9) tran(nn,1),tran(nn,2),tran(nn,3)
  22                  continue
                  endif
                  close (unit=9)
                  dfreq=tran(2,1)-tran(1,1)
                  write(ilog,*)'QQ dfreq =', dfreq
c
c        do wave analysis
          elseif (iopt.eq.4) then
             goto 200
c
c        store results
          elseif (iopt.eq.8) then
                  write(*,*)'STORE'
                  write(*,*)'Choose:  0=return, 1=store asc, 2=unf, 3=bin'
                  read(*,*) istore
                  write(ilog,*) istore,' ::istore'
                  if (istore.eq.0) goto 900
                  write(*,'(a\)')' type:  output filename--> '
                  read(5,'(1a30)') file2
```

```
             write(ilog,*) file2,' ::filename'
             call fylopn(file2,istore)
             if (istore.eq.1) then
                do 810 nn=1,ndata
                   freq=dfreq*real(nn-1)
                   write(9,*) freq,tran(nn,2),tran(nn,3)
810             continue
             else
                do 820 nn=1,ndata
                   freq=dfreq*real(nn-1)
                   write(9) freq,tran(nn,2),tran(nn,3)
820             continue
             endif
             close (unit=9)
          endif
          goto 900
c
200          continue
          write(*,*)'     '
          write(*,*)'        WAVES menu:         '
          write(*,*)'          40 : return      '
          write(*,*)'          41 : strain from force        '
          write(*,*)'          42 : viscoelastic       '
          write(*,*)'          43 : dispersion       '
          write(*,*)'          44 : reflected from elastic boundary '
          write(*,*)'          45 :       ..       .. concentrated mass'
          write(*,*)'          46 : reflected from elastic joint     '
          write(*,*)'          47 : transmitted at      ..      ..   '
          write(*,'(a\)')' select one--> '
          read(5,*) iwav
          write(ilog,*) iwav,' ::WAVES'
c
c          strain from force
          if (iwav.eq.41) then
             c0=sqrt(e1/rho1)
             area = thick1*width1
             ea=e1*area
c
c          viscoelastic
          elseif (iwav.eq.42) then
             area1 = thick1*width1
             eta=eta1
c
c          dispersion
          elseif (iwav.eq.43) then
             c0=sqrt(e1/rho1)
             znu=znu1
             area1 = thick1*width1
             rgyr = sqrt(thick1*thick1/6.0)
c
c          reflected from spring
          elseif (iwav.eq.44) then
             c0=sqrt(e1/rho1)
             area = thick1*width1
             ea=e1*area
             bigk=e2
c
c          reflected from concentrated mass
          elseif (iwav.eq.45) then
             c0=sqrt(e1/rho1)
```

```
                area = thick1*width1
                ea=e1*area
c
c          reflected from joint
           elseif (iwav.eq.46) then
                c0=sqrt(e1/rho1)
                c02=sqrt(e2/rho2)
                area1 = thick1*width1
                area2 = thick2*width2
                ea=e1*area1
                ea2=e2*area2
                r2=(ea2/c02)/(ea/c0)
                r3=1.
c
c          transmitted from joint
           elseif (iwav.eq.47) then
                c0=sqrt(e1/rho1)
                c02=sqrt(e2/rho2)
                area1 = thick1*width1
                area2 = thick2*width2
                ea=e1*area1
                ea2=e2*area2
                r2=(ea2/c02)/(ea/c0)
                r3=1.
c
           endif
c
c          BIG DO_LOOP OVER FREQUENCIES
           do 30 nn=2,nyqst
                omega = tran(nn,1)*2.0*pi*fscale
                beta=omega*sqrt(rho1/e1)
c
c               simple transmission
                if (iwav.eq.41) then
                     arg=beta*xx2
                     tf2=-cexp(-ci*arg)/ea
                     tff=-cexp(-ci*0)/ea
c                    obtain strain
                     fcmplx = tran(nn,2)+ci*tran(nn,3)
                     ecmplx = fcmplx*tf2
                     ocmplx = ecmplx*1000000.
c
c               viscoelastic transmission
                elseif (iwav.eq.42) then
c                    viscoelastic modulus
                     eomega=e1*(e2+ci*omega*eta)/(e1+e2+ci*omega*eta)
                     xk=omega*csqrt(rho1/eomega)
                     arg=ci*xk*xx2
                     if (cabs(real(arg)).gt.40) then
                          tf2=0.0
                     else
                          tf2=+cexp(-ci*xk*xx2)
                          tff=+cexp(-ci*xk*xx2)
                     endif
c                    obtain strain
                     fcmplx = tran(nn,2)+ci*tran(nn,3)
                     ecmplx = fcmplx*tf2
                     ocmplx = ecmplx*1.00
c
c               dispersion transmission
```

```
                  elseif (iwav.eq.43) then
                     disp=sqrt(1+(znu*rgyr*omega/c0)**2)
                     xk=omega/c0*disp
                     tf2=-cexp(-ci*xk*xx2)/e1/area1
                     tff=-cexp(-ci*xk*xx2)/e1/area1
c                    obtain strain
                     fcmplx = tran(nn,2)+ci*tran(nn,3)
                     ecmplx = fcmplx*tf2
                     ocmplx = ecmplx*1000000.
c
c                 elastic spring boundary
                  elseif (iwav.eq.44) then
                     xk=omega/c0
                     tf2=(ci*xk*ea-bigk)/(ci*xk*ea+bigk)
                     tf2=tf2*cexp(ci*xk*(xx2-xbnd))
                     tff=(ci*xk*ea-bigk)/(ci*xk*ea+bigk)
c                    obtain force
                     fcmplx = tran(nn,2)+ci*tran(nn,3)
                     ecmplx = fcmplx*tf2
                     ocmplx =-ecmplx
c
c                 concentrated mass
                  elseif (iwav.eq.45) then
                     xk=omega/c0
                     tf1=(-zmass*omega**2 )/(2*ci*xk*ea+zmass*omega**2)
                     tf2=(2*ci*xk*ea     )/(2*ci*xk*ea+zmass*omega**2)
                     tff=cabs(tf1)+ci*cabs(tf2)
c                    otain force
                     fcmplx = tran(nn,2)+ci*tran(nn,3)
                     ecmplx = fcmplx*tf2
                     ocmplx =-ecmplx
c
c                 elastic joint
                  elseif (iwav.eq.46 .or. iwav.eq.47) then
                     beta=omega/c0
                     beta2=omega/c02
                     beta3=omega/c0
                     cex2m= cexp(-ci*beta2*blen)
                     cex2p= cexp(+ci*beta2*blen)
                     cex3m= cexp(-ci*beta3*blen)
c
c                    elastic joint matrix   B1  A2  B2  A3
                     a(1,1)= -1
                     a(1,2)= 1
                     a(1,3)= 1
                     a(1,4)= 0
                         a(2,1)= 1
                         a(2,2)= r2
                         a(2,3)=-r2
                         a(2,4)= 0
                     a(3,1)= 0
                     a(3,2)= cex2m
                     a(3,3)= cex2p
                     a(3,4)=-cex3m
                         a(4,1)= 0
                         a(4,2)=-r2*cex2m
                         a(4,3)= r2*cex2p
                         a(4,4)= r3*cex3m
                     b(1)=1
                     b(2)=1
```

```
                   b(3)=0
                   b(4)=0
                   call det(a,b,d)
                   tf1 = d(1)/d(5)
                   tf2 = d(4)/d(5)
c
c                  obtain reflected
                   if (iwav.eq.46) then
                       fcmplx = tran(nn,2)+ci*tran(nn,3)
                       ecmplx = fcmplx*tf1
                       ocmplx = ecmplx
                       tff=tf1
c                  obtain transmitted
                   elseif (iwav.eq.47) then
                       fcmplx = tran(nn,2)+ci*tran(nn,3)
                       ecmplx =-fcmplx*tf2*cex3m
                       ocmplx = ecmplx
                       tff=tf2
                   endif
                 endif
c
c                FORM THE COMPLEX CONJUGATE
                 tran(nn,1)          = tran(nn,1)*fscale
                 tran(nn,2)          = real(ocmplx)
                 tran(nn,3)          = aimag(ocmplx)
                 tran(2+ndata-nn,2)  = tran(nn,2)
                 tran(2+ndata-nn,3)  =-tran(nn,3)
 30        continue
c
c          ASSIGN ZERO TO INITIAL VALUE
           sum=0.0
           do 32 n=2,nyqst
              sum=sum+tran(n,2)
 32        continue
           tran(1,2)          = -2*sum
           tran(1,3)          = 0.0
           tran(2+ndata-1,2) = tran(1,2)
           tran(2+ndata-1,3) =-tran(1,3)
           goto 900
c
  999      stop'stopped from menu'
           end
c
c
           subroutine fylopn(fyl,iopen)
             character*30 fyl
             if (iopen.eq.1) then
                 open(9,file=fyl)
             elseif (iopen.eq.2) then
                 open(9,file=fyl,form='unformatted')
             elseif (iopen.eq.2) then
                 open(9,file=fyl,form='binary')
             endif
             rewind 9
             return
             end
c
c
        subroutine det(a,b,d)
          complex d11,d22,d33,d44
```

```
        complex a(4,4),b(4),c(4),d(5)
c
        d11 = a(1,1)*( a(2,2)*(a(3,3)*a(4,4)-a(4,3)*a(3,4))
     >                -a(2,3)*(a(3,2)*a(4,4)-a(4,2)*a(3,4))
     >                +a(2,4)*(a(3,2)*a(4,3)-a(4,2)*a(3,3)))
        d22 =-a(1,2)*( a(2,1)*(a(3,3)*a(4,4)-a(4,3)*a(3,4))
     >                -a(2,3)*(a(3,1)*a(4,4)-a(4,1)*a(3,4))
     >                +a(2,4)*(a(3,1)*a(4,3)-a(4,1)*a(3,3)))
        d33 = a(1,3)*( a(2,1)*(a(3,2)*a(4,4)-a(4,2)*a(3,4))
     >                -a(2,2)*(a(3,1)*a(4,4)-a(4,1)*a(3,4))
     >                +a(2,4)*(a(3,1)*a(4,2)-a(4,1)*a(3,2)))
        d44 =-a(1,4)*( a(2,1)*(a(3,2)*a(4,3)-a(4,2)*a(3,3))
     >                -a(2,2)*(a(3,1)*a(4,3)-a(4,1)*a(3,3))
     >                +a(2,3)*(a(3,1)*a(4,2)-a(4,1)*a(3,2)))
        d(5) = d11 + d22 + d33 + d44
        do 30 m=1,4
          do 20 n=1,4
            c(n)=a(n,m)
            a(n,m)=b(n)
 20       continue
c
c
        d11 = a(1,1)*( a(2,2)*(a(3,3)*a(4,4)-a(4,3)*a(3,4))
     >                -a(2,3)*(a(3,2)*a(4,4)-a(4,2)*a(3,4))
     >                +a(2,4)*(a(3,2)*a(4,3)-a(4,2)*a(3,3)))
        d22 =-a(1,2)*( a(2,1)*(a(3,3)*a(4,4)-a(4,3)*a(3,4))
     >                -a(2,3)*(a(3,1)*a(4,4)-a(4,1)*a(3,4))
     >                +a(2,4)*(a(3,1)*a(4,3)-a(4,1)*a(3,3)))
        d33 = a(1,3)*( a(2,1)*(a(3,2)*a(4,4)-a(4,2)*a(3,4))
     >                -a(2,2)*(a(3,1)*a(4,4)-a(4,1)*a(3,4))
     >                +a(2,4)*(a(3,1)*a(4,2)-a(4,1)*a(3,2)))
        d44 =-a(1,4)*( a(2,1)*(a(3,2)*a(4,3)-a(4,2)*a(3,3))
     >                -a(2,2)*(a(3,1)*a(4,3)-a(4,1)*a(3,3))
     >                +a(2,3)*(a(3,1)*a(4,2)-a(4,1)*a(3,2)))
        d(m) = d11 + d22 + d33 + d44
c
          do 22 n=1,4
            a(n,m)=c(n)
 22       continue
 30     continue
c
        return
        end
```

ROOTCOMP

```
c       ROOTCOMP.for : obtain complex roots.
c
        dimension wave(1100,3)
        complex fxy, zz, ci,xk
        character*30 file2
          ci=cmplx(0,1.)
          pi=4.0*atan(1.0)
c
        write(*,'(///)')
        write(*,*)' ROOTROOTROOTROOTROOTROOTROOTROOTROOTROOTROOTROOT'
        write(*,*)'              version 1.05, May 1988'
        write(*,*)'              ikayex SOFTWARE TOOLS'
        write(*,*)' ROOTROOTROOTROOTROOTROOTROOTROOTROOTROOTROOTROOT'
```

```
          write(*,'(///)')
c
          write(*,*)' # of iters  |  step size  |  resolution'
          read(*,*) iter,d, f00
          write(*,*)'Choose:  0=no print  1=print'
          read(*,*) iprint
          write(*,*)' Input:  curved beam parameter  xi'
          read(*,*) zi
          write(*,*)' Input:  initial freq | dfreq | # of freqs'
          read(*,*) w0,dw,npt
          write(*,*)' Input initial guess: xkx0  |  xky0'
          read(*,*) xkx0,xky0
               xk=xkx0+ci*xky0
               zz=xk*xk
               x0=real(zz)
               y0=aimag(zz)
               write(*,*)' initial guess for zz =',x0,y0
c
               dkeep=d
c        loop over frequency range
          do 40 n=1,npt
               omb=w0+dw*(n-1)
               d=dkeep
               zz=x0+ci*y0
               call ffn(zz,fxy,omb,zi)
               f9=cabs(fxy)
c             iteration loop
               do 20 i=1,iter
                    zz=x0+ci*y0
                    call ffn(zz,fxy,omb,zi)
                    f0=cabs(fxy)
                    if(f0.gt.f9) d=d/2
                    if(f0.lt.f00) goto 30
c
c              neighboring function evaluations
                    zz=x0+d+ci*(y0)
                    call ffn(zz,fxy,omb,zi)
                    f1=cabs(fxy)
                    zz=x0   +ci*(y0+d)
                    call ffn(zz,fxy,omb,zi)
                    f2=cabs(fxy)
c
c                estimates of slopes
                    df1=f1-f0
                    df2=f2-f0
                    df3=sqrt(df1*df1+df2*df2)
                    x0=x0-d*df1/df3
                    y0=y0-d*df2/df3
                    f9=f0
c
                    if (iprint.eq.1) then
                        write(*,*) x0,y0,f0
                        write(*,*)' fs',f1,f2
                    endif
  20           continue
c
  30           continue
               zz=x0+ci*y0
               xk=csqrt(zz)
               if (iprint.eq.2) then
```

```
              write(*,*)' iteration =',i,' zz =',zz
           endif
           wave(n,1)=omb/(2*pi)
           wave(n,2)=real(xk)
           wave(n,3)=abs(aimag(xk))
40         continue
c
           read(*,'(1a30)') file2
           open(unit=2,file=file2)
           write(*,*)' writing <text> file'
           do 50 n=1,npt
              write(2,*) wave(n,1),wave(n,2),wave(n,3)
50         continue
c
999        continue
           stop' ok'
           end
c
c
           subroutine ffn(z,f,omb,zi)
           complex z,f
c
              omb2=omb*omb
              zi2=zi*zi
              a0 = 12*omb2*(-zi2+omb2)
              a1 = -zi2*zi2 - omb2*(12-zi2)
              a2 = -(0*2*zi2+omb2)
              f=a0 + z*(a1 + z*(a2 + z))
c
           return
           end
c
c    SAMPLE DRIVER FILE
c    100 .011 1e-6                    :: iter  d   f00
c    2                                :: 1=print
c    0.30                             :: zi
c    0.61  -.005   121                :: w0   dw   npt
c    1.70  0.0                        :: x0   y0
c    s411.2                           :: filename
```

FORCCOMP

```
c          FORCCOMP.for : determine force history due to impact.
c
           dimension fi(1030),vsi(1030),xout(1030,6)
           character*30 file3
           pi=4.0*atan(1.0)
           ilog=67
           open(unit=ilog,file='force.log')
           rewind ilog
c
           write(*,'(//)')
           write(*,*)'FORCEFORCEFORCEFORCEFORCEFORCEFORCEFORCEFORCEFORCE'
           write(*,*)'             version 3.03, August 1988'
           write(*,*)'             ikayex SOFTWARE TOOLS'
           write(*,*)'FORCEFORCEFORCEFORCEFORCEFORCEFORCEFORCEFORCEFORCE'
           write(*,'(////)')
1          continue
           write(*,'(//)')
```

```
        write(*,*)' MAIN menu :'
        write(*,*)'         0 :  quit'
        write(*,*)'         1 :  input plate data'
        write(*,*)'         2 :  concentrated mass'
        write(*,*)'         3 :  rod'
        write(*,*)'         4 :  beam'
        write(*,*)'         5 :  plate'
        write(*,*)'         8 :  store results'
        write(*,*)'Choose --> '
        read(*,*) iopt
        write(ilog,*) iopt,' ::MAIN'
c
        if (iopt.eq.0) goto 999
        if (iopt.eq.1) goto 100
        if (iopt.eq.2) goto 1000
        if (iopt.eq.3) goto 1000
        if (iopt.eq.4) goto 1000
        if (iopt.eq.5) goto 1000
        if (iopt.eq.8) goto 800
           goto 1
c
c       read matl properties
 100    continue
        read(*,*) eb,znub,rhob
           write(ilog,*) eb,znub,rhob,' ::obj    e nu rho'
        read(*,*) dim1,dim2
           write(ilog,*) dim1,dim2 ,' ::obj    dim1 dim2'
        read(*,*) es,znus,rhos
           write(ilog,*) es,znus,rhos,' ::ball    e nu rho'
        read(*,*) rads
           write(ilog,*) rads,' ::ball    rad  '
        read(*,*) velzero
           write(ilog,*) velzero,' ::initial vel '
        read(*,*) isteps,dt,timezero
           write(ilog,*) isteps,dt,timezero,' ::# of steps, tinc tzero'
        read(*,*) iecho,imetric
           write(ilog,*) iecho,imetric,' :: echo=1  metric=1'
c
c          CONSTANTS
c          conc. mass
                vol2=pi*(dim1**2)*dim2
                zmass2=vol2*rhob
                radc=1000.0
c       rods
                arear=pi*(dim1**2)
                c0=sqrt(eb/rhob)
                coeffr = c0/eb/arear
c       beams
                thick=dim1
                width=dim2
                areab=width*thick
                zi=1./12.*width*thick**3
                ei=eb*zi
                ra=rhob*areab
                   coef1=thick/(4.*ei*sqrt(2*pi))*(ei/ra)**.25
                   coef2=sqrt(ei/ra)*2./thick
                   coeffb=coef1*coef2
c       plates
                thick=dim1
                dd = eb*thick**3/12/(1-znub**2)
```

```
                      coeffp = 1.0/8.0/sqrt(rhob*thick*dd)
c           sphere
                      vols=4./3.*pi*rads**3
                      zmass=rhos*vols
c
c           CONTACT LAWS
c           hertzian
                      zkb=eb/(1-znup**2)
                      zks=es/(1-znus**2)
                      zk = 4./3.*sqrt(rads*radc/(rads+radc))*zkb*zks/(zkb+zks)
c           plastic rod
                      tau=eb*areab/zmass/c0
                      taur=1.0/tau
c           plastic beam
                      coef3=1.0/(zmass*coef1*coef2*pi)
                      tau=8.0*sqrt(rhob*thick*ei/dim2)/zmass
                      taub=1.0/tau
c           plastic plate
                      tau=8.0*sqrt(rhob*thick*dd)/zmass
                      taup=1.0/tau
c
            goto 1
c
c           BIG LOOP
 1000       continue
c
c           Initial Conditions
            time=timezero
            us0=0.0
            usdot0=velzero
            usdotdot0=0.0
            u20=0.0
            u2dot0=0.0
            u2dotdot0=0.0
c
            fi(0)=0.
            fi(1)=0.
            vsi(0)=velzero
            vsi(1)=velzero
c
            write(*,*)'PRINT menu:'
            write(*,*)'         0: no echo   '
            write(*,*)'         1: time force'
            write(*,*)'         2: time force strain'
            write(*,*)'         3: time elastic plastic'
            write(*,*)'         4: time strain obj_disp vel'
            write(*,*)'         5: all'
            write(*,*)'Choose-->'
            read(*,*) iprint
            write(ilog,*)iprint, ' ::PRINT'
c
            if (iprint.eq.0) then
c              goto 22
            elseif (iprint.eq.1) then
               write(*,*) time,force
            elseif (iprint.eq.2) then
               write(*,*) time,force,strain
            elseif (iprint.eq.3) then
               write(*,*) time,force,fplastic
            elseif (iprint.eq.4) then
```

```
                write(*,*) time,strain,u20,usdot0
            elseif (iprint.eq.5) then
                print*,0.0 ,force,strain,u20,usdot0,fplastic
            endif
c
c           SIMPLE TRAPEZOIDAL  ALGORITHM
            do 20  i=1,isteps
                time=timezero+dt*real(i-2)
                if (i.lt.3) goto 17
c
                us   = us0 + usdot0*dt
                usdot=usdot0 + usdotdot0*dt
                u2   = u20 + u2dot0*dt
                    difzz = us-u2
                    if(difzz.lt.0)  difzz=0
                usdotdot0 =-zk/zmass*difzz**1.5
c
c               concentrated mass
                if (iopt.eq.2) then
                    u2dot=u2dot0 + u2dotdot0*dt
                    u2dotdot0 = zk/zmass2*difzz**1.5
                    u2dot0=u2dot
                    fplastic=velzero*zmass/taur*exp(-(time-timezero)/taur)
c
c               rods
                elseif (iopt.eq.3) then
                    u2dot0 = coeffr*zk*difzz**1.5
                    strain =-u2dot0/c0*1000000.
                    fplastic=velzero*zmass/taur*exp(-(time-timezero)/taur)
c
c               beams
                elseif (iopt.eq.4) then
                    fi(i)=difzz
                    vsi(i)=u2
                    call intgl(time,dt,i,fi,zintgl,vsi,zintv)
                    u2dot0 =-coeffb*zk*zintgl
                    strain = sqrt(3.0)/c0*u2dot0*1e6
                    fplastic=velzero*zmass/taub*exp(-(time-timezero)/taub)
c
c               plate
                elseif (iopt.eq.5) then
                    u2dot0 = coeffp*zk*difzz**1.5
                    strain = sqrt(3.0)/c0*u2dot0*1e6
                    fplastic=velzero*zmass/taup*exp(-(time-timezero)/taup)
c
                endif
c
                us0=us
                usdot0=usdot
                u20=u2
                force = zk*difzz**1.5
c
   17           continue
                if (imetric.eq.1) then
                    force=force*4.4482
                    fplastic=fplastic*4.4482
                    u20=u20*25.4
                    usdot0=usdot0*.0254
                endif
c
```

```
          xout(i,1)=time
          xout(i,2)=force
          xout(i,3)=strain
          xout(i,4)=u20
          xout(i,5)=usdot0
          xout(i,6)=fplastic
c
      if (iprint.eq.0) then
c         goto 20
      elseif (iprint.eq.1) then
          write(*,*) time,force
      elseif (iprint.eq.2) then
          write(*,*) time,force,strain
      elseif (iprint.eq.3) then
          write(*,*) time,force,fplastic
      elseif (iprint.eq.4) then
          write(*,*) time,strain,u20,usdot0
      elseif (iprint.eq.5) then
          write(*,*) time,force,strain,u20,usdot0,fplastic
      endif
c
20        continue
          xout(1,1)=0.0
          xout(1,2)=0.0
              if (imetric.eq.1) then
                  zmass=zmass*.45359
                  velzero=velzero*0.0254
                  usdot=usdot*.0254
              endif
      zmom1=zmass*velzero
      zmom2=zmass*usdot
      zmomdf=zmom1-zmom2
      write(*,*)' momentum: init  fin  diff',zmom1,zmom2,zmomdf
c
      goto 1
c
c         STORE RESULTS
800       continue
          write(*,*)' type :  output filename'
          read(*,'(1a30)') file3
          write(ilog,*) file3,' ::filename'
          open(unit=3,file=file3)
          rewind 3
          write(*,*)' type :  print selection'
          read(*,*) iprint
          write(ilog,*) iprint,' ::print'
            write(*,*) file3,iprint
c
          do 820 i=1,isteps
            if (iprint.eq.1) then
              write(3,*) xout(i,1),xout(i,2)
            elseif (iprint.eq.2) then
              write(3,*) xout(i,1),xout(i,2),xout(i,3)
            elseif (iprint.eq.3) then
              write(3,*) xout(i,1),xout(i,2),xout(i,6)
            elseif (iprint.eq.4) then
              write(3,*) xout(i,1),xout(i,3),xout(i,4),xout(i,5)
            elseif (iprint.eq.5) then
                write(3,*) (xout(i,j),j=1,6)
            endif
```

```
820       continue
          close (unit=3)
          goto 1
c
 999      continue
          stop'stopped from menu'
          end
c
c
          subroutine intgl(t,tinc,inmax,fi,zintgl,vsi,zintv)
          dimension fi(501),vsi(501)
c
          zintv=0.0
          zinteg=0.
          do 31 in=1,inmax
             tau1=real(in-0)*tinc
             difft=(abs(t-tau1))**.5-(abs(t-tau1+tinc))**.5
             fav=(fi(in-1)+fi(in))/2.
             if (in.eq.1) fav=fi(1)/2.
             fav=fav**1.5
             zinteg = zinteg + 2.0*fav*difft
             vsav=(vsi(in-1)+vsi(in))/2.
             zintv = zintv + 2.0*vsav*difft
   31     continue
          zintgl=zinteg
          return
          end
c
c         SAMPLE DRIVER FILE
c         1                                      ::MAIN
c         450000.000000  4.000000E-01  1.140000E-04  ::obj  e nu rho
c         1.000000       2.500000E-01               ::obj  dim1 dim2
c         3.000000E+07   3.000000E-01  7.340000E-04  ::ball e nu rho
c         5.000000E-01                              ::ball    rad
c         124.000000                                ::initial vel
c         100            1.000000E-06  0.000000E+00  ::# steps  tinc t0
c         1   0                                     ::echo=1 metric=1
c         4                                         ::MAIN
c         1                                         ::PRINT
c         8                                         ::MAIN
c         rtemp                                     ::filename
c         1                                         ::print
c         0                                         ::MAIN
```

References

[1] Abbas, B.A.H. and Thomas, J., 1977, "The Second Frequency Spectrum of Timoshenko Beams," Journal of Sound and Vibration, Vol. 51, pp. 123-137.

[2] Abbot, B.W. and Broutman, L.J., 1966, "Stress-wave Propagation in Composite Materials," Experimental Mechanics, Vol. 6, pp. 383-384.

[3] Abdelhamid, M.K. and McConnell, K.G., 1986, "A Spectral-Analysis Method for Nonstationary Field Measurements," Experimental Mechanics, Vol. 26, pp. 47-55.

[4] Abou-Sayed, A.S., Clifton, R.J., and Hermann, L., 1976, "The Oblique-plate Impact Experiment," Experimental Mechanics, Vol. 16, pp. 127-132.

[5] Abramowitz, M., and Stegun, I. A., 1965, Handbook of Mathematical Functions, Dover, New York.

[6] Abramson, H.N., 1957, "Flexural Waves in Elastic Beams of Circular Cross Section," The Journal of the Acoustical Society of America, Vol. 29(1), pp. 42-46.

[7] Abramson, H.N., Plass, H.J., and Ripperger, E.A., 1958, "Stress Wave Propagation in Rods and Beams," Advances in Applied Mechanics, Vol.5, pp.111-194.

[8] Achenbach, J. D., 1973, Wave Propagation in Elastic Solids, North-Holland, New York.

[9] Ambartsumyan, S.A., 1970, Theory of Anisotropic Plates, Technomic, Stamford, Conn..

[10] Aprahamian, R., Evensen, D.A., Mixson, J.S., and Wright, J.E., 1971, "Application of Pulsed Holographic Interferometry to the Measurement of Propagating Transverse Waves in Beams," Experimental Mechanics, Vol. 11, pp. 309-314.

[11] Aprahamian, R., Evensen, D.A., Mixson, J.S., and Jacoby, J.L., 1971, "Holographic Study of Propagating Transverse Waves in Plates," Experimental Mechanics, Vol. 11, pp. 357-362.

[12] Atkins, K.J. and Hunter, S.C., 1975, "The Propagation of Longitudinal Elastic Waves around Right- Angled Corners in Rods of Square Cross-Sections," Quarterly Journal of Mech. appl. Math., Vol.XXVIII, pp. 245-260.

[13] Austin, A.L., 1965, "Measurements of Thermally Induced Stress Waves in a Thin Rod Using Birefringent Coatings," Experimental Mechanics, Vol. 5, pp. 1-10.

[14] Bache, T.C. and Hegemier, G.A., 1974, "On Higher-Order Elastodynamic Plate Theories," Journal of Applied Mechanics, Vol. 41, pp. 423-428.

[15] Bahar, L.Y., and Hetnarski, R.B., 1979, "Direct Approach to Thermoelasticity," Journal of Thermal Stresses, vol 2(1), pp. 135-147.

[16] Barez, F., Goldsmith, W., and Sackman, J.L., 1980, "Longitudinal Wave Propagation in Axisymmetric Structures with Material and/or Aereal Discontinuity," Experimental Mechanics, Vol. 20, pp. 325-333.

[17] Barton, C.S., and Volterra, E.G., and Citron, S.J., 1958, "On Elastic Impacts of Spheres on Long Rods," Proceedings of third Cong. Applied Mechanics, pp.89-94.

[18] Bergland, G.D., 1969, "A Guided Tour of the Fast Fourier Transform," IEEE Spectrum, Vol 6, pp. 41-52.

[19] Bhashyam, G.R. and Prathap, G.P., 1981, "The Second Frequency Spectrum of Timoshenko Beams," Journal of Sound and Vibration, Vol. 76, pp. 407-420.

[20] Bellanger, M., 1984, *Digital Processing of Signals*, Wiley, New York.

[21] Blackman, R.B., and Tukey, J.W., 1958, *The Measurement of Power Spectra*, Dover, New York.

[22] Bickle, L.W., 1970, "The Response of Strain Gages to Longitudinally Sweeping Strain Pulses," Experimental Mechanics, Vol. 10, pp. 333-337.

[23] Bingham, C., Godfrey, M.D. and Tukey, J.W., 1967, "Modern Techniques of Power Spectrum Estimation," IEEE Transactions on Audio and Electroacoustics, Vol. Au-15, pp. 56-66.

[24] Boley, B.A., 1955, "An Approximate Theory of Lateral Impact on Beams," Journal of Applied Mechanics, Vol. 21-22, pp. 69-76.

[25] Boley, B.A. and Chao, C.C., 1955, "Some Solutions of the Timoshenko Beam Equations," Journal of Applied Mechanics, Vol. 22, pp. 579-586.

[26] Boley, B.A. and Chao, C-C., 1958, "An Approximate Analysis of Timoshenko Beams Under Dynamic Loads," Journal of Applied Mechanics, Vol. 25, pp. 31-36.

[27] Booer, A.K., Chambers, J., and Mason, L.M., 1977, "Fast Numerical Algorithm for the Recompression of Dispersed Time Signals," Electronic Letters, Vol.13, pp. 453-455.

[28] Bowman, F., 1958, **Introduction to Bessel Functions**, Dover, New York.

[29] Brazier-Smith, P.R., Butler, D. and Halstead, J.R, 1981, "The Determination of Propagation Path Lengths of Dispersive Flexural Waves Through Structures," Journal of Sound and Vibration, Vol. 75, pp. 453-457.

[30] Brigham, E.O., 1973, *The Fast Fourier Transform*, Prentice-Hall, Englewood Cliffs, New Jersey.

[31] Britton, W.G.B. and Langley, G.O., 1968, "Stress Pulse Dispersion in Curved Mechanical Waveguides," Journal of Sound and Vibration, Vol. 7(3), pp. 417-430.

[32] Chambers. Ll.G., 1962, "Sommerfeld Type Diffraction Problems for a Plate," Zeitschrift fur Angewandte Math. u. Mech. B. 42, s. 545-555.

[33] Chatfield, C., 1984, *The Analysis of Time Series: An Introduction*, Chapman and Hall, London.

[34] Chow, T.S., 1971, "On the Propagation of Flexural Waves in an Orthotropic Laminated Plate and Its Response to an Impulsive Load," Journal of Composite Materials, Vol. 5, pp 306-319.

[35] Clough, R.W. and Penzien, J., 1975, *Dynamics of Structures*, McGraw-Hill, New York.

[36] Cochran, W.T., Cooley, J.W., Favin, D.L., Helms, H.D., Kaenel, R.A., Lang, W.W., Maling, G.C. Jr., Nelson, D.E., Rader, C.M., and Welch, P.D., 1967, "What is the Fast Fourier Transform?," IEEE Transactions of Audio and Electroacoustics, Vol. Au-15, pp. 45-55.

[37] Conway, H.D. and Jakubowski, M., 1969, "Axial Impact of Short Cylindrical Bars," Journal of Applied Mechanics, Vol. 36. pp. 809-813.

[38] Cooley, J.W. and Tukey, J.W., 1965, "An Algorithm for the Machine Calculation of Complex Fourier Series," Math.Comput., Vol. 19, pp. 297-301.

[39] Cooley, J.W., Lewis, P.A.W. and Welch, P.D., 1967, "Application of the Fast Fourier Transform to Computation of Fourier Integrals, Fourier Series, and Convolution Integrals," IEEE Transactions of Audio and Electroacoustics, Vol. Au-15, pp. 79-84.

[40] Cooley, J.W., Lewis, P.A.W. and Welch, P.D., 1970, "The Fast Fourier Transform Algorithm: Programming Considerations in the Calculation of Sine, Cosine and Laplace Transforms," Journal of Sound and Vibration, Vol. 12, pp. 315-337.

[41] Cornelius, W.K. and Kubitza, W.K., 1970, "Experimental Investigation of Longitudinal Wave Propagation in an Elastic Rod with Coulomb Friction," Experimental Mechanics, Vol. 10, pp. 137-144.

[42] Cowper, G.R., 1966, "The Shear Coefficient in Timoshenko's Beam Theory," Journal of Applied Mechanics, Vol. 33, pp. 335-340.

[43] Crowley, F.B., III, Phillips, J.W., and Taylor, C.E., 1972, "Pulse Propagation in Straight and Curved Beams-Theory and Experiment," Journal of Applied Mechanics, Vol. 39, pp. 1-6.

[44] Cunningham, D.M. and Goldsmith, W., 1958, "Short-Time Impulses Produced by Longitudinal Impact," S.E.S.A. Proceedings, Vol. XVI No. 2, pp. 153-162. and Ripperger, E.A., 1958, Discussion, pp. 163-165.

[45] Curtis, C.W., 1960, Propagation of an Elastic Strain Pulse in A semi-infinite bar, in "Stress Wave Propagation in Materials," (Ed. N. Davids) p. 15-43. Interscience, New York.

[46] Daniel, I.M., Liber, T., and LaBedz, R.H., 1979, "Wave Propagation in Transversely Impacted Composite Laminates," Experimental Mechanics, Vol. 19, pp. 9-16.

[47] Das, A. and Roy, S.K., 1979, "Note on the Forced Vibration of an Orthotropic Plate on an Elastic Foundation," Journal of Sound and Vibration, Vol. 66, 521-525.

[48] Dauer, F.W., 1978, "Experimental Determination of Impact Force vs. Displacement Curves," Experimental Mechanics, Vol. 18, pp. 159-160.

[49] Davies, R.M., 1948, "A Critical Study of the Hopkinson Pressure Bar," Philosophical Transactions of the Royal Society, Vol. 240, pp. 375-457.

[50] Davies, E.D.H. and Hunter, S.C., 1963, "The Dynamic Compression Testing of Solids by the Method of the Split Hopkinson Pressure Bar," Phys. Solids, Vol. 11, pp. 155-179.

[51] Davids, N., 1959, "Transient Analysis of Stress-Wave Penetration in Plates," Journal of Applied Mechanics, Vol. 26, pp. 651-660.

[52] Dawe, D.J., 1978, "Finite Strip Models for Vibration of Mindlin Plates," Journal of Sound and Vibration, Vol. 59, pp. 441-452.

[53] Dengler, M.A. and Goland, M., 1952, "Transverse Impact of Long Beams, Including Rotatory Inertia and Shear Effects," Proceedings of First U.S. National Congress of Applied Mechanics, pp. 179-186.

[54] Desmond, T.P., 1981, "Theoretical and Experimental Investigation of Stress Waves at a Junction of Three Bars," Journal of Applied Mechanics, Vol. 48, pp. 148-154.

[55] DeVault, G.P., and Curtis, C.W., 1962, "Elastic Cylinder with Free Lateral Surface and Mixed Time-Dependent End Conditions," The Journal of the Acoustical Society of America, Vol. 34(4), pp. 421-432.

[56] Doyle, J.F., 1984a, "An Experimental Method for Determining the Dynamic Contact Law,"Experimental Mechanics, Vol. 24, pp. 10-16.

[57] Doyle, J.F., 1984b, "Further Developments in Determining the Dynamic Contact Law," Experimental Mechanics, Vol. 24, pp. 265-270.

[58] Doyle, J.F., and Kamle, S., 1985a, "An Experimental Study of the Reflection and Transmission of Flexural Waves at Discontinuities," Journal of Applied Mechanics, Vol. 52, pp. 699-673.

[59] Doyle, J.F., and Kamle, S., 1985b, "An Experimental Study of the Reflection and Transmission of Flexural Waves at a T-Joint," Proceeding for Society of Experimental Mechanics, Las Vegas, pp. 560-564.

[60] Doyle, J.F., 1986a, "Determining the Contact Force During the Transverse Impact of Plates," Proceedings of Society for Experimental Mechanics, New Orleans, pp. 560-567.

[61] Doyle, J.F., 1986b, "Digital Waveform Recorders for Dynamic Strain Measurements," Proceedings of Society for Experimental Mechanics, New Orleans, pp. 74-78.

[62] Doyle, J.F., 1986d, "Toward in Situ Testing of the Mechanical Properties of Composite Panels," Proceedings of Society for Experimental Mechanics, Keystone, Colo., pp.165-172.

[63] Doyle. J.F., 1986e, "Application of the Fast Fourier Transform (FFT) to Wave Propagation Problems," Int. Journal of Analytical and Experimental Modal Analysis, Vol. 1, pp. 18-25.

[64] Doyle, J.F., and Kamle, S., 1987a, "An Experimental Study of the Reflection and Transmission of Flexural Waves at an Arbitrary T-Joint," Journal of Applied Mechanics, vol. 54, pp. 136-140.

[65] Doyle, J.F., 1987b, "Determining the Contact Force During the Transverse Impact of Plates," Experimental Mechanics, Vol. 27, pp. 68-72.

[66] Doyle, J.F., 1987c, "An Experimental Method for Determining the Location and Time of Initiation of an Unknown Dispersing Pulse," Experimental Mechanics, Vol. 27, pp. 229-233.

[67] Doyle, J.F., 1987d, "Experimentally Determining the Contact Force During the Transverse Impact of Orthotropic Plates," Journal of Sound and Vibration, Vol. 118, pp. 441-448.

[68] Doyle, J.F., 1987e, "Transverse Impact of a Plate near a Straight edge or Stiffener," Proceedings of Society of Experimental Mechanics, Houston. pp. 917-923.

[69] Doyle, J.F.,1987f, "A Spectrally Formulated Finite Element for Longitudinal Wave Propagation", Int. Journal of Analytical and Experimental Modal Analysis, Vol. 3, pp. 1-5.

[70] Doyle, J.F., Farris, T.N. and Ledington, H.M.1987g, "The Interaction of Flexural Waves with a Delamination Crack", Int. Conf. on Photomechanics and Speckle Metrology, San Diego. pp. 665-661.

[71] Doyle, J.F., 1988a, "Toward In-Situ Testing of the Mechanical Properties of Composite Panels," Journal of Composite Materials, vol 22, pp. 416-426.

[72] Doyle, J.F., 1988b, "Spectral Analysis of Coupled Thermoelastic Waves," Journal of Thermal Stresses, vol 11, pp. 175-185.

[73] Doyle, J.F., Farris, T.N. and Ledington, H.M., 1988c, "The Interaction of Flexural Waves with a Lengthwise Delamination Crack", Proceedings of Society for Experimental Mechanics, Portland, pp. 1253-1260.

[74] Doyle, J.F. and Farris, T.N., 1989a, "Wave Propagation in a Split Timoshenko Beam," to appear in Journal of Sound and Vibration.

[75] Doyle, J.F. and Farris, T.N., 1989b, "A Spectrally Formulated Finite Element for Longitudinal Wave Propagation", submitted to Int. Journal of Analytical and Experimental Modal Analysis.

[76] Duffey, T.A. and Key, S.W., 1969, "Experimental-Theoretical Correlations of Impulsively Loaded Clamped Circular Plates," Experimental Mechanics, Vol. 9, pp. 241-249.

[77] Dyer, I., 1960, "Moment Impedance of Plates," The Journal of the Acoustical Society of America, Vol.32, pp. 1290-1297.

[78] Efron, L. and Malvern, L.E., 1969, "Electromagnetic Velocity-transducer Studies of Plastic Waves in Aluminum Bars," Experimental Mechanics, Vol. 9, pp. 255-262.

[79] Elmore, W.C. and Heald, M.A., 1985, *Physics of Waves*, Dover, New York.

[80] Eubanks, R.A., Muster, D., and Volterra, E., 1954, "On the Attenuation of a Sinusoidal Pulse in a Cylindrical Specimen Held Between Elastic Bars," Proceedings of Second U.S. National Congress of Applied Mechanics, pp. 193-200.

[81] Ewing, W.M., and Jardetzky, W.S., 1957, *Elastic Waves in Layered Media*, McGraw-Hill, New York.

[82] Farrar, C.L., 1984, "Impact Response of a Circular Membrane," Experimental Mechanics, Vol. 24, pp. 144-149.

[83] Felix, M.P., 1977, "Determination of Stress Levels for Dynamic Fractures of Oil Shale," Experimental Mechanics, Vol. 17, pp. 381-384.

[84] Folk, R., Fox, G., Shook, C.A., and Curtis, C.W., 1958, "Elastic Strain Produced by Sudden Application of Pressure to One End of a Cylindrical Bar.I. Theory," The Journal of the Acoustical Society of America, Vol. 30(6) pp. 552-558.

[85] Galletly, G.D., 1959, "Circular Plates on a Generalized Foundation," Journal of Applied Mechanics, Vol. 26, p. 297.

[86] Ganesan, N. and Dhotarad, M.S., 1983, "Vibration Analysis of Mindlin Plates," Journal of Sound and Vibration, Vol. 87, pp. 643-645.

[87] Gazis, D.C., and Mindlin, R.D., 1960, "Extensional Vibrations and Waves in a Circular Disk and a Semi-Infinite Plate," Journal of Applied Mechanics, Vol. 27, pp. 541-547.

[88] Gere, J.M. and Lin, Y.K., 1958, "Coupled Vibrations of Thin-Walled Beams of Open Cross Section," Journal of Applied Mechanics, Vol. 25, pp. 373-378.

[89] Goel, R.P., 1976, "Transverse Vibrations of Tapered Beams," Journal of Sound and Vibration, Vol. 47, pp. 1-7.

[90] Goland, M., Wickersham, P.D. and Dengler, M.A., 1955, "Propagation of Elastic Impact in Beams in Bending," Journal of Applied Mechanics, Vol. 21-22, pp. 1-7.

[91] Goldsmith, W., 1960, *Impact*, Edward Arnold LTD., London.

[92] Goldsmith, W. and Cunningham, D.M., 1955, "An Experimental Investigation of the Oblique Impact of Spheres Upon Simply Supported Steel Beams," S.E.S.A. Proceedings, Vol. XIV, pp.171-180.

[93] Goldsmith, W., Liu, T.W., and Chulay, S., 1965, "Plate Impact and Perforation by Projectiles," Experimental Mechanics, Vol. 5, pp. 385-404.

[94] Goodman, L.E., 1952, "Circular-Crested Vibrations of an Elastic Solid Bounded by Two Parallel Planes," Proceedings of First U.S. National Congress of Applied Mechanics, pp. 65-73.

[95] Graff, K.F, 1970, "Elastic Wave Propagation in a Curved Sonic Transmission Line," IEEE Transactions on Sonics and Ultrasonics, Vol. SU-17, pp. 1-7.

[96] Gupta, B., 1979,"Propagation of Elastic Waves in Rods With Variable Cross Section," Journal of Applied Mechanics, Vol. 46, p. 951.

[97] Gupta, B.V.R., Naarayanan, S. and Ganesan, N., 1986, "Free Vibration Characteristics of Periodically Stiffened Panels with Damped Stringers," Journal of Sound and Vibration, Vol. 105(2), pp. 351-356.

[98] Haines, D.W., 1974, "Approximate Theories for Wave Propagation and Vibrations in Elastic Rings and Helical Coils of Small Pitch," Int. Journal of Solids and Structures, Vol. 10, pp. 1405-1416.

[99] Haines, D.W. and Chang, N., 1975, "The Helix as a Mechanical Filter - An Example of Continuous Mechanical Systems Analyzed by use of the Fourier Time Transform and Complex Frequency Response," Mechanics Research Communications, Vol. 2, Nos. 5-6, pp. 245-247.

[100] Haines, D.W. and Chang, N., 1976, "Transmission and Reflection of Extensional Stress Pulses at the Junction of a Straight Bar and a Helix," Journal of Applied Mechanics, Vol. 43, pp. 309-313.

[101] Hayashi, T., Ugo, R., and Morimoto, Y., 1986, "Experimental Observation of Stress Waves Propagating in Laminated Composites," Experimental Mechanics, Vol. 26, pp. 169-174.

[102] Harting, D.R., 1972, "Digital Transient-test Techniques," Experimental Mechanics, Vol. 17, pp. 335-340.

[103] Hasan, S.A. and Barr, A.D.S., 1974, "Linear Vibration of Thin-Walled Beams of Equal Angle-Section," Journal of Sound and Vibration, Vol. 32, pp. 3-23.

[104] Hsieh, D.Y., and Kolsky, H., 1958, "An Experimental Study of Pulse Propagation in Elastic Cylinders," Proceedings of Physical Society, Vol. 71, pp. 608-612.

[105] Hsieh, D.Y. and Lee, J.P., 1973, "Experimental Study of Pulse Propagation in Curved Elastic Rods," The Journal of the Acoustical Society of America, Vol. 54, No. 4, pp. 1052-1055.

[106] Ho, C.L., 1975, "Ultra-sonic Surface Eave Detection Technique," Chap. 3, Experimental Techniques in Fracture Mechanics, 2, ed. A.S. Kobayashi, Iowa State Press.

[107] Holmes, P.J., 1974, "The Experimental Characterization of Wave Propagation Systems: II. Continuous Systems and the effects of Dispersion," Journal of Sound and Vibration, Vo. 35, pp. 277-297.

[108] Ince, E.L., 1956, *Ordinary Differential Equations*, Dover, New York.

[109] Jahsman, W.E.,1971, "Reexamination of the Kolsky Technique for Measuring Dynamic Material Behavior," Journal of Applied Mechanics, Vol. 38, pp. 1-8.

[110] Jensen, J.J., 1983, "On the Shear Coefficient in Timoshenko's Beam Theory," Journal of Sound and Vibration, Vol. 87, pp. 621-636.

[111] Jones, A.T., 1970, "Vibration of Beams Immersed in a Liquid," Experimental Mechanics, Vol. 10, pp. 84-88.

[112] Jones, R.P.N., 1954, "The Wave Method for solving Flexural Vibration Problems," Journal of Applied Mechanics, Vol. 21, pp. 75-80.

[113] Jones, R.M., 1975, *Mechanics of Composite Materials*, McGraw-Hill, New York.

[114] Jordan, R.W. and Whiston, G.S., 1984, "Remote Impact Analysis by Use of Propagated Acceleration Signals, II: Comparison Between Theory and Experiment," Journal of Sound and Vibration, Vol. 97(1) pp. 53-63.

[115] Kalnins, A., 1966, "On Fundamental Solutions and Green's Functions in the Theory of Elastic Plates," Journal of Applied Mechanics, Vol. 33, pp. 31-38.

[116] Kandianis, F., 1973,"Correlation Techniques in the Analysis of Transient Processes," Journal of Sound and Vibration, Vol. 26, pp. 161-172.

[117] Kane, T.R., 1954, "Reflection of Flexural Waves at the Edge of a Plate," Journal of Applied Mechanics, Vol. 21-22, pp. 213-220.

[118] Katsamanis, F. and Goldsmith, W., 1982, "Fluid Effects and Response in Transverse Impact on Liquid-filled Tubes," Experimental Mechanics, Vol. 22, pp. 245-255.

[119] Kaul, R.K. and McCoy, J.J., 1964, "Propagation of Axisymmetric Waves in a Circular Semiinfinite Elastic Rod," The Journal of the Acoustical Society of America, Vol. 36(4), pp. 653-660.

[120] Kennedy, L.W. and Jones, O.E., 1969, "Longitudinal Wave Propagation in a Circular Bar Loaded Suddenly by a Radially Distributed End Stress," Journal of Applied Mechanics, Vol. 36, pp. 470-478.

[121] Kenner, V.H. and Goldsmith, W., 1968, "Elastic Waves in Truncated Cones," Experimental Mechanics, Vol. 2, pp. 442-449.

[122] Kenner, V.H., 1980, "The Fluid Hopkinson Bar," Experimental Mechanics, Vol. 20, pp. 226 -232.

[123] Kenner, V.H., Staab, G.H., and Jing, H.S., 1984, "A Study of Flaw Identification in Adhesive Bonds Using a Technique of Impact Modification," Experimental Mechanics, Vol. 24, pp. 243-247.

[124] Kenner, V.H., 1986, "Elastic-Pulse Propagation in Thin Hollow Cones," Experimental Mechanics, Vol. 26, pp. 192-193.

[125] Kida, S. and Oda, J., 1982, "On Fracture Behavior of Britle Cantilever Beam Subjected to Lateral Impact Load," Experimental Mechanics, Vol. 22, pp. 69-74.

[126] Klyukin, I.I., and Sergeev, Yu.D., 1964, "Scattering of Flexural Waves by Antivibrators on a Plate," Soviet Physics-Acoustics (English trans. of Akusticheskli Zhurnal), Vol. 10, pp. 49-53.

[127] Knauss, W.G., 1968, "Uniaxial Wave Propagation in a Viscoelastic Material Using Measured Materials Properties," Journal of Applied Mechanics, Vol. 35, pp. 449-453.

[128] Kolsky, H., 1963, *Stress Waves in Solids*, Dover, New York.

[129] Konenkov, Y.K., 1964, "Diffraction of a Flexural Wave by a Circular Obstacle in a Plate," Soviet Physics-Acoustics, Vol. 10,No.2, pp. 153-156.

[130] Krylov, V.I., and Skoblya, N.S., 1977, *A Handbook of Methods of Approximate Fourier Transformation and Inversion of the Laplace Transformation*, Mir Publishers, Moscow.

[131] Kunukkasseril, V.X., and Chandrasekharan, K., 1975, "Concentrated-impact Loading of Circular Plates," Experimental Mechanics, Vol. 15, pp. 424-428..

[132] Kuo, S.S., 1961, "Beam Subjected to Eccentric Longitudinal Impact," Experimental Mechanics, Vol. 11, pp. 102-108.

[133] Lamb, H., 1904, "On the Propagation of Tremors over the Surfaces of an Elastic Solid," Philosophical Transactions of the Royal Society of London, A 203, pp. 1-42.

[134] Lee, E.H., 1940, "The Impact of a Mass Striking a Beam," Journal of Applied Mechanics, Vol. 7, pp. A-129-A-138.

[135] Lee, J.P. and Kolsky, H., 1972, "The Generation of Stress Pulses at the Junction of Two Noncollinear Rods," Journal of Applied Mechanics, Vol.15, pp. 809-813.

[136] Lee, T.M. and Sechler, E.E., 1975, "Longitudinal Waves in Wedges," Experimental Mechanics, Vol. 9, pp. 41-48.

[137] Levinson, M., 1981, "A New Rectangular Beam Theory," Journal of Sound and Vibration, Vol. 74, pp. 81-87.

[138] Lewis, J.L., Goldsmith, W., and Cunningham, D.M., 1969, "Internal-strain Measurements of Longitudinal Pulses in Conical Bars," Experimental Mechanics, Vol. 22, pp. 313-320.

[139] Lifshitz, J.M. and Mor, G., 1982, "Elastic-pulse Propagation in Thin Hollow Cones," Experimental Mechanics, Vol. 22, pp. 166-170.

[140] Lindholm, U.S. and Yeakley, L.M., 1968, "High Strain-rate Testing: Tension and Compression," Experimental Mechanics, Vol. 8, pp. 1-9.

[141] Ljunggren, S., 1983, "Generation of Waves in an Elastic Plate by a Vertical Force and by a Moment in the Vertical Plane," Journal of Sound and Vibration, Vol. 90, pp. 559-584.

[142] Love, A.E., 1944, A Treatise on the Mathematical Theory of Elasticity, Dover, New York.

[143] Lyon, R.L. and Zable, J.L., 1973, "Impact-force Source and Impact-force Calibrator," Experimental Mechanics, Vol. 13, pp. 257-264.

[144] Mace, B.R., 1984, "Wave Reflection and Transmission in Beams," Journal of Sound and Vibration, Vol. 97(2), pp. 237-246.

[145] Maiden, C.J., 1958, "The Stresses Produced in a Thin Elastic Plate by a Transverse Impulsive Force," Phil. Mag., Vol. 3, pp. 1413-1423.

[146] Maginness, M.G., 1972, "The Reconstruction of Elastic Wave Fields from Measurements over a Transducer Array," Journal of Sound and Vibration, Vol. 20 (2), pp. 219-240.

[147] Mandel, J.A., Mathur, R.J., and Chang, Y.C., 1971, "Stress Waves at Rigid Right Angle Joint," Journal of the Engineering Mechanics Division, ASCE, Vol. 97, pp. 1173-1186.

[148] Martin, H.C.,1966, Introduction to Matrix Methods of Structural Analysis, McGraw-Hill.

[149] McConnell, K.G. and Chang, C.N., "A Study of the Axial-Torsional Coupling Effect on a Sagged Transmission Line," Experimental Mechanics, Vol. 26, pp. 324-329.

[150] Mead, D.J., 1982, "A Comparison of Some Equations for the Flexural Vibration of Damped Sandwich Beams," Journal of Sound and Vibration, Vol. 83, pp. 363- 377.

[151] Mead, D.J., 1986, "A New Method of Analyzing Wave Propagation in Periodic Structures; Applications to Periodic Timoshenko Beams and Stiffened Plates," Journal of Sound and Vibration, Vol. 104(1), pp. 9-27.

[152] Medick, M.A., 1961, "On Classical Plate Theory and Wave Propagation," Journal of Applied Mechanics, Vol. 28, pp. 223-228.

[153] Miklowitz, J., 1953, "Flexural Wave Solutions of Coupled Equation Representing the More Exact Theory of Bending," Journal of Applied Mechanics, Vol. 19-20, pp. 511-514.

[154] Miklowitz,J., 1955, "Traveling Compressional Waves in an Elastic Rod According to the More Exact One-Dimensional Theory," Proceeding of Second National U.S. Congress of Applied Mechanics, pp. 179-186.

[155] Miklowitz, J., 1957, "The Propagation of Compressional Waves in a Dispersive Elastic Rod, Part I-Results from the theory," Journal of Applied Mechanics, Vol. 24, pp. 231-239.

[156] Miklowitz, J. and Nisewanger, C.R., 1957, "The Propagation of Compressed Waves in a Dispersive Elastic Rod, Part II-Experimental results and comparison with theory," Journal of Applied Mechanics, Vol. 24, pp. 240-244.

[157] Miklowitz, J., 1958, "On the Use of Approximate Theories of an Elastic Rod in Problems of Longitudinal Impact," Proceedings of Third U.S. National Congress of Applied Mechanics, pp. 215-224.

[158] Miklowitz, J., 1960, "Flexural Stress Waves in an Infinite Elastic Plate Due to a Suddenly Applied Concentrated Transverse Load," Journal of Applied Mechanics, Vol. 27, pp. 681-689.

[159] Mindlin, R.D., 1951, "Influence of Rotatory Inertia and Shear on Flexural Motions of Isotropic, Elastic Plates," Journal of Applied Mechanics, Vol. 18, pp. 31-38.

[160] Mindlin, R.D., 1958, "Vibrations of an Infinite, Elastic Plate at its Cut-Off Frequencies," Proceedings of Third U.S. National Congress on Applied Mechanics, pp. 225-226.

[161] Mindlin, R.D., and McNiven, H.D., 1960, "Axially Symmetric Waves in Elastic Rods," Journal of Applied Mechanics, Vol. 27, pp. 145-151.

[162] Mindlin, R.D. and Herrmann G., 1950, "A One-Dimensional Theory of Compressional Wave in an Elastic Rod," Proceedings of First U.S. National Congress of Applied Mechanics, pp. 187-191

[163] Mindlin, R.D. and Deresiewicz, H.D., 1954, "Timoshenko's Shear Coefficient for Flexural Vibrations of Beams," Proceedings of Second National U.S. Congress of Applied Mechanics, pp. 175-178.

[164] Miles, A.W., 1976, "Shock-front Loading Method for Studies in Dynamic Photoelasticity," Experimental Mechanics, Vol. 76, pp. 349-355.

[165] Morely, S.D., 1961, "Elastic Waves in a Naturally Curved Rod," Quarterly Journal of Mechanical and Applied Mathematics, Vol. 14, pp. 155-172.

[166] Mori, D., 1957, "Lateral Impact on Bars and Plates," S.E.S.A. Proceedings, Vol. XV, pp. 171-178.

[167] Mortimer, R.W., Rose, J.L., and Chou, P.C., 1972, "Longitudinal Impact of Cylindrical Shells," Experimental Mechanics, Vol. 12, pp. 25-31.

[168] Nakao, T., Okano, T., and Asano, I., 1985, "Theoretical and Experimental Analysis of Flexural Vibrations of the Viscoelastic Timoshenko Beam," Journal of Applied Mechanics, Vol. 52. pp. 728-731.

[169] Narayanan. S. and Shanbhag, R.L., 1982, "Sound Transmission Through a Damped Sandwich Panel," Journal of Sound and Vibration, Vol. 80, pp. 315-327.

[170] Nevill, Jr.,G.E., and Hoese, F.O., 1965, "Impulsive Loading Using Sprayed Silver Acetylide-Silver Nitrate," Experimental Mechanics, Vol. 5, pp. 294-298.

[171] Nevill, Jr., G.E., Sierakowski, C.A., Ross, C.A., and Jones, E.R., 1972, "One-dimensional Wave Pulses in Steel-Epoxy Composites," Experimental Mechanics, Vol. 12, pp. 278-282.

[172] Noll, A.M., 1964, "Short-Time Spectrum and "Cepstrum" Techniques for Vocal-Pitch Detection," The Journal of the Acoustical Society of America, Vol. 36 No. 2, pp. 296-302.

[173] Norman, T., 1986, "Transverse Impact of Stiffened Laminated Composite Panels," M.S. Thesis, Purdue University.

[174] Oi, K., 1966, "Transient Response of Bonded Strain Gages," Experimental Mechanics, Vol. 6, pp. 463-469.

[175] Okada, A., Cunningham, D.M., and Goldsmith, W., 1968, "Stress Waves in Pyramids by Photoelasticity," Experimental Mechanics, Vol. 8, pp. 289-299.

[176] Oliver, J., 1957, "Elastic Wave Dispersion in a Cylindrical Rod by a Wide-Band Short-Duration Pulse Technique," The Journal of the Acoustical Society of America, Vol. 29. pp. 189-194.

[177] Onoe, M., McNiven, H.D., and Mindlin, R.D., 1962, "Dispersion of Axially Symmetric Waves on Elastic Rods," Journal of Applied Mechanics, Vol. 29, pp. 729-734.

[178] Pao, Y.-H., and Mindlin, R.D., 1960, "Dispersion of Flexural Waves in an Elastic, Circular Cylinder," Journal of Applied Mechanics, Vol. 27, pp. 513-520.

[179] Papoulis, A., 1962, *The Fourier Integral and its Applications*, McGraw-Hill, New York.

[180] Pavic, G. and White, R.G., 1977, "On the Determination of Transmission Path Importance in Dispersive Systems," Acustica, Vol. 38, pp. 76-80.

[181] Percival, C.M. and Cheney, J.A., 1969, "Thermally Generated Stress Waves in a Dispersive Elastic Rod," Experimental Mechanics, Vol. 9, pp. 49-57.

[182] Philipson, L.L., 1956, "On the Role of Extension in the Flexural Vibrations of Rings," Journal of Applied Mechanics, Vol. 23, pp. 364-366.

[183] Plass, H.J., 1958, "Some Solutions of the Timshenko Beam Equation for Short Pulse-Type Loading," Journal of Applied Mechanics, Vol. 25, pp. 379-385.

[184] Press, W.H., Flannery, B.P., Teukolsky, S.A., and Vetterling, W.T., 1986, *Numerical Recipes*, Cambridge University Press, Cambridge.

[185] Przemieniecki, J.S.,1986, *Theory of Matrix Structural Analysis*, Dover, New York.

[186] Rader, D. and Mao, M., 1972, "Amplification of Longitudinal Stress Pulses in Elastic Bars with an Intermediate Tapered Region," Experimental Mechanics, Vol. 12, pp. 90-94.

[187] Reddy, G.N. and Saha, S., 1983, "A Highly Sensitive Noncontacting Electromagnetic Device for Detecting Dynamic Stress Structure," Experimental Mechanics, Vol. 23, pp. 418-424.

[188] Reismann, H. and Pawlik, P.S., 1974, *Elastokinetics*, West Publishing Co., New York.

[189] Reismann, H., 1959, "Forced Vibrations of a Circular Plate," Journal of Applied Mechanics, Vol. 26, pp. 526-527.

[190] Reismann, H., and Yamaguchi, T., 1981, "Wave Motion in Non-Homogeneous Beams and Plates-A Comparison of Two Theories," SM Archives, Vol. 6, Issue 2, pp. 213-277.

[191] Schell, A.C., Chairman of Editing Committee, 1979, Programs for Digital Signal Processing, IEEE Press.

[192] Ripperger, E.A., 1953, "Propagating Pulses in Cylindrical Bars — An Experimental Study," First Midwestern Conf. on Solid Mechanics, pp. 29-39.

[193] Ripperger, E.A. and Abramson, N.H., 1957, "A Study of Propagation of Flexural Waves in Elastic Beams," Journal of Applied Mechanics, Vol. 24, pp. 431- 434.

[194] Ripperger, E.A. and Yeakley, L.M., 1963, "Measurement of Particle Velocities Associated with Waves Propagating in Bars," Experimental Mechanics, Vol. 3, pp. 47-56.

[195] Ripperger, E.A. and Abramson, H.N., 1957, "Reflection and Transmission of Elastic Pulses in a Bar at a Discontinuity in Cross Section," Third Midwestern Conf. on Solid Mechanics, pp. 29-39.

[196] Rizzi, S.A., and Doyle, J.F., 1987a, "Use of Whole-field Data to Characterize In-Plane Stress Waves," Society for Experimental Mechanics, Spring meeting, Houston, pp. 864-869.

[197] Rizzi, S.A., and Doyle, J.F., 1987b, "Determination of Dynamic Bending Stiffness of an Orthotropic Cross-Ply Composite," Mechanical Systems and Signal Processing, Vol. 1, pp. 285-291.

[198] Rizzi, S.A., and Doyle, J.F., 1988, "Experimental Study of the Impact of a Surface with a Soft Layer," Proceedings Society for Experimental Mechanics, Porland, pp. 1213-1220.

[199] Rose, J.L., Mortimer, R.W., and Blum, A., 1973, "Elastic-wave Propagation in a Joined Cylindrical-Conical-Cylindrical Shell," Experimental Mechanics, Vol. 13, pp. 150-156.

[200] Roy, A.K. and Plunkett, R., 1985, "Dispersive Bending Waves in Uniform Bars," Experimental Mechanics, Vol. 25, pp. 308-311.

[201] Schwieger, H., 1970, "Central Deflection of a Transversely Struck Beam," Experimental Mechanics, Vol. 10, pp. 166-169.

[202] Seelig, J.M. and Hoppmann, W.H.II, 1964, "Impact on an Elastically Connected Double-Beam System," Journal of Applied Mechanics, Vol. 31, pp. 621-626.

[203] Shoua, E.D., 1968, "The Composite Damping Capacity of Sandwich Cantilever Beams," Experimental Mechanics, Vol. 8, pp. 300-308.

[204] Smith, E.A.L., 1955, "Impact and Longitudinal Wave Transmission," J. of Applied Mechanics, Vol. 22, pp. 963-973.

[205] Sneddon, I.N., 1951, *Fourier Transforms*, McGraw-Hill, New York.

[206] Soedel, W., 1981, *Vibrations of Shells and Plates*, Marcel Dekker, New York.

[207] Speigel, M.R., 1974, *Fourier Analysis with Applications to Boundary Value Problems*, McGraw-Hill, New York.

[208] Srawley, J.E. and Esgor, J.P., 1966, "Investigation of Hydrotest Failure of Thiokol Chemical Corporation 260-inch-diam er SL-1 Motor Case," NASA TMX-1194.

[209] Stephen, N.G., 1978, "On e Variation of Timoshenko's Shear Coefficient With Frequency," Journal of Applied Mechan cs, Vol. 45, pp. 695-697.

[210] Stephen, N.G. and Levins n, M., 1979, "A Second Order Beam Theory," Journal of Sound and Vibration, Vol. 67, pp 293-305.

[211] Stephen, N.G., 1982, "The Second Frequency Spectrum of Timoshenko Beams," Journal of Sound and Vibration, Vol. 80, pp. 578-582.

[212] Stephenson, J.G. and Wilhoit, J.C. Jr., 1965, "An Experimental Study of Bending Impact Waves in Beams," Experimental Mechanics, Vol. 5, pp. 16-21.

[213] Suh, N.P, 1970, "Helical Coils as Impact Load Dispersers," Journal of Engineering for Industry, Vol. 92, Series B, No.1, pp. 197-207.

[214] Sun, C.T., 1971, "Microstructure Theory for a Composite Beam," Journal of Applied Mechanics, Vol. 38, pp. 947-954.

[215] Sun, C.-T., 1973, "Propagation of Shock Waves in Anisotropic Composite Plates," Journal of Composites Materials, Vol. 7. pp. 366-382.

[216] Suzuki, S.-i., 1986, "Axisymmetric Dynamic Behaviour of Thick Plates Subjected to Impulsive Loads," Journal of Sound and Vibration, Vol 105(2), pp. 339-345.

[217] Thorkildsen, R.L. and Hoppmann, W.H.II, 1959, "Effect of Rotatory Inertia on the Frequencies of Vibration of Stiffened Plates," Journal of Applied Mechanics, Vol. 26, pp. 298-300.

[218] Timoshenko, S., and Woinowsky-Krieger, S., 1959, *Theory of Plates and Shells*, McGraw-Hill, New York.

[219] Tribolet, J.M. and Quartieri, T.F., 1970, "Computation of the Complex Cepstrum," Chap. 7, Programs for Digital Signal Processing, IEEE Press.

[220] Truell, R., Elbaum, C., and Chick B.B., 1969, *Ultrasonic Methods in Solid State Physics*, Academic Press, New York.

[221] Vasudeva, R.Y. and Bhaskara, R.K., 1978, "Propagation of a Pulse in Nonhomogeneous Elastic Rod With Varying Cross Section," Vol. 45, pp. 942-944.

[222] Vigness, I., 1949, "Transverse Waves in Beams," S.E.S.A. Proceedings, Vol. xx, pp. 69-81.

[223] Viktorov, I.A., 1967, **Rayleigh and Lamb Waves Physical Theory and Applications**, Plenum Press, New York.

[224] Vold, H., Crowley, J., and Rocklin, G.T., 1984, "New Ways of Estimating Frequency Response Functions," Journal of Sound and Vibration, Vol. 97, pp. 34-38.

[225] Volterra, E., 1958, "On the Response to External Loads of Continuous Elastic Systems with Hereditary Damping Characteristics," Proceedings of Third U.S. National Congress of Applied Mechanics, pp. 57-63.

[226] Walter, P.L. and Nelson, H.D., 1979, "Limitations and Corrections in Measuring Structural Dynamics," Experimental Mechanics, Vol. 19, pp. 309-316.

[227] Walter, P.L., 1981, "Decovolution as a Technique to Improve Measurement-system Data Integrity," Experimental Mechanics, Vol. 21, pp. 309-314.

[228] Wang, J.T.S., Liu, Y.Y. and Gibby, J.A., 1983, "Vibrations of Split Beams," Journal of Sound and Vibration, Vol. 84, pp. 491-502.

[229] White, P.H., 1969, "Cross Correlation in Structural Systems: Dispersion and Nondispersion Waves," Journal of the Acoustical Society of America, Vo. 45, pp. 1118-1128.

[230] Whiston, G.S., 1984, "Remote Impact Analysis by Use of Propagated Acceleration Signals, I: Theoretical Methods," Journal of Sound and Vibration, Vol. 97(1), pp. 35-51.

[231] Wilshaw, T.R. and Kelly, J.M., 1968, "Response of Circular Clamped Plates to Square-wave Stress Pulses," Experimental Mechanics, Vol. 8, pp. 450-458.

[232] Witrick, W.H., 1966, "On Elastic Wave Propagation in Helical Springs," International Journal of Mechanical Science, Vol. 8, pp. 25-47.

[233] Yang, T.Y.,1986, *Finite Element Structural Analysis*, Prentice-Hall.

[234] Yang, J.C.S. and Hassett, R.J., 1972, "Transient Stresses in Axisymmetric Bodies of Varying Area," Experimental Mechanics, Vol. 12, pp. 304-310.

[235] Yong, K.H. and Atkins, K.J., 1983, "Generation of Elastic Stress Waves at a T-Junction of Square Rods," Journal of Sound and Vibration, Vol. 88, pp. 431-436.

[236] Zachmanoglou, E.C., and Volterra, E., 1958, "An Engineering Theory of Longitudinal Wave Propagation in Cylindrical Elastic Rods," Third U.S. Cong. of Applied Mechanics, pp. 239-245.

[237] Zachary, L.W. and Burger, C.P., 1980, "Dynamic Wave Propagation in a Single Lap Joint," Experimental Mechanics, Vol. 20, pp. 162-166.

[238] Zemanek, J. (Jr) and Rudnick, I., 1961, "Attenuation and Dispersion of Elastic Waves in a Cylindrical Bar," The Journal of the Acoustical Society of America, Vol. 33, pp. 1283-1288.

ADDITIONAL REFERENCES

REFERENCES: Related to two-dimensional problems

[239] Burger, C.P. and Riley, W.F., 1974, "Effects of Impedance Mismatch on the Strength of Waves in Layered Solids," Experimental Mechanics, Vol. 14, pp. 129-137.

[240] Christie, D.G., 1955, "Reflections of Elastic Waves from a Free Boundary," Philosophical Mag., series 7, Vol. 46, pp. 527-541.

[241] Daniel, I.M. and Marino, R.L., 1971, "Wave Propagation in Layered Model due to Point-source Loading in Low-impedance Medium," Experimental Mechanics, Vol. 11, pp. 210-216.

[242] Eason, G., 1964, "On the Torsional Impulsive Loading of an Elastic Half Space," Journal of Mechanical and Applied Math, Vol. 17, pp. 279-292.

[243] Flynn, P.D., 1973, "Dual-beam Polariscope and Framing Camera for Dynamic Photoelasticity," Experimental Mechanics, Vol. 13, pp. 1-7.

[244] Freund, L.B. and Phillips, J.W., 1970, "Near-Front Stress Singularity for Impact on an Elastic Quarter-Space," The Journal of the Acoustical Society of America, Vol. 47, No. 3,(part 2), pp. 942-943.

[245] Gilbert, F. and Knopoff, L., 1959, "Scattering of Impulsive Elastic Waves by a Rigid Cylinder," The Journal of the Acoustical Society of America, Vol. 31, pp. 1169-1175.

[246] Goodier, J.N., and Bishop, R.E.D., 1952, "A Note on Critical Reflections of Elastic Waves at Free Surfaces," Journal of Applied Physics, Vol. 23, pp. 124-126.

[247] Goodier, J.N., Jahsman, W.E., and Ripperger, E.A., 1959, "An Experimental Surface-Wave Method for Recording Force-Time Curves in Elastic Impacts," Journal of Applied Mechanics, Vol. 26, pp. 3-7.

[248] Gutenberg, B., 1944, "Energy Ratio of Reflected and Refracted Seismic Waves," Bull. seism. Soc. Am., Vol. 34, pp. 85-102.

[249] Holloway, D.C., Patacca, A.M., and Fourney, W.L., 1977, "Application of Holographic Interferometry to a Study of Wave Propagation in Rock," Experimental Mechanics, Vol. 77, pp. 281-289.

[250] Hunter, S.C., 1957, "Energy Absorbed by Elastic Waves During Impact," Journal Mechanical Physics and Solids, Vol. 5, pp. 162-171.

[251] Kawata, K., and Hashimoto, S., 1967, "On Some Differences between Dynamic- and Static-stress Distributions," Experimental Mechanics, Vol. 7, pp. 91-96.

[252] Kawata, K., Takeda, N., and Hashimoto, S., 1984, "Photoelastic-coating Analysis of Dynamic Stress Concentration in Composite Strips," Experimental Mechanics, Vol. 24, pp. 316-327.

[253] Ko, W.L. and Karlsson, T., 1967, "Application of Kirchhoff's Integral Equation Formulation to an Elastic Wave Scattering Problem," Journal of Applied Mechanics, Vol. 34, pp. 921-930.

[254] Kuske, A., 1966, "Photoelastic Research on Dynamic Stresses," Experimental Mechanics, Vol. 6, pp. 105-112.

[255] Kuske, A., 1977, "Photoelastic Stress Analysis of Machines Under Dynamic Load," Experimental Mechanics, Vo. 17, pp. 88-96.

[256] Lewis, D. and Dally, J.W., 1970, "Photoelastic Analysis of Rayleigh Wave Propagation in Wedges," Journal of Geophysical Research, Vol. 75 (17), pp. 3387-3398.

[257] McCoy, J.J., 1968, "Effects of Non-Propagating Plate Waves on Dynamical Stress Concentrations," International Journal of Solids and Structures, Vol. 4, pp. 355-370.

[258] Miklowitz, J., 1969, "Analysis of Elastic Waveguides Involving an Edge," *Wave Propagation in Solids*, ASME, New York, (editor), pp.44-70.

[259] Miles, J.W., 1960, "Homogeneous Solutions in Elastic Wave Propagation," Quarterly of Applied Mathematics, Vol. 18, No. 1, pp. 37-59.

[260] Miller, G.F. and Pursey, H., 1954, "The Field and Radiation Impedance of Mechanical radiators on the Free Surface of a Semi-infinite Isotropic Solid," Proceedings of the Royal Society, A 223, pp. 521-541.

[261] Miller, G.F. and Pursey, H., 1955, "On the Partiton of Energy Between Elastic Waves in a Semi-infinite Solid," Proceedings of the Royal Society, A 233, pp. 55-69.

[262] Mitra, M., 1964, "Disturbance Produced in an Elastic Half-space by Impulsive Normal Pressure," Proceedings of Cambridge Philosophical Society,Math, Vol. 60, pp. 683-696.

[263] Moon, F.C., 1972, "Wave Surfaces Due to Impact on Anisotropic Plates," Journal of Composite Materials, Vol. 6, pp. 62-79.

[264] Moon, F.C., 1973, "Stress Wave Calculations in Composite Plates using the Fast Fourier Transform," Computers and Structures, Vol.3, pp. 1195-1204.

[265] Pao, Y-H, 1962, "Dynamical Stress Concentration in an Elastic Plate," Journal of Applied Mechanics, Vol. 29, pp. 299-305.

[266] Pao, Y-H. and Chao, C.C., 1964, "Diffractions of Flexural Waves by a Cavity in an Elastic Plate," AIAA Journal, Vol. 2, pp 2004-2010.

[267] Pao, Y.H., and Mow, C.C., 1973, *The Diffraction of Elastic Waves and Dynamic Stress Concentrations*, Crane, Russak and Company Inc., New York.

[268] Pekeris, C.L., 1955a, "The Seismic Surface Pulse," Proceedings of the National Acad. Sci., U.S.A., Vol. 41, pp. 469-480.

[269] Pekeris, C.L., 1955b, "The Seismic Buried Pulse," Proc. Natn. Acad. Sci, Vol 41, pp. 629-638.

[270] Pekeris, C.L. and Lifson, H., 1957, "Motion of the Surface of a Uniform Elastic Half-Space Produced by a Buried Pulse," Journal of the Acoustical Society of America, Vol. 29, pp. 1233-1238.

[271] Pih, H. and Snyders, L.S., 1969, "Photoelastic Studies of Ultrasonic Waves in a Large Plate," Experimental Mechanics, Vol. 9, pp. 186-192.

[272] Riley, W.F. and Dally, J.W., 1969, "Recording Dynamic Fringe Patterns with a Cranz-Schardin Camera," Experimental Mechanics, Vol. 9, pp. 27N-33N.

[273] Stump, B.W., 1984, "Stress Waves in an Elastic Half-Space: Single and Multiple Surface Source," Vol. 92, pp. 181-201.

[274] Tatel, H.E., 1954, "Note of the Nature of a Seismogram -II," Journal of Geophysical Research, Vol. 59,(2), pp. 280-294.

[275] Thau, S.A. and Dally, J.W., 1969, "Subsurface Characteristics of the Rayleigh Wave," International Journal of Engineering Science, Vol. 7, pp. 37-52.

[276] Vasudevan, N., and Mal, A.K., 1985, "Responses of an Elastic Plate to Localized Transient Sources," Journal of Applied Mechanics, Vol. 52. pp. 356-362.

[277] White, R.M., 1958, "Elastic Wave Scattering at a Cylindrical Discontinuity in a Solid," The Journal of the Acoustical Society of America, Vol. 30, pp. 771-785.

[278] Williams, D.D. and Smith, D.G., 1974, "A Simple Way to Make Plane Waves for Dynamic Photoelasticity," Experimental Mechanics, Vol. 14, pp. 19N-20N.

REFERENCES: Related to exact solutions

[279] Fox, G. and Curtis, C.W., 1958, "Elastic Strain Produced by Sudden Application of Pressure to One End of a Cylindrical Bar. II. Experimental Observations," The Journal of the Acoustical Society of America, Vol. 30, pp. 559-563.

[280] Fraser, W.B. 1970, "Longitudinal Elastic Waves in Square Bars," Journal of Applied Mechanics, Vol. 37, pp. 537-538.

[281] Gilbert, F. and Knopoff, L., 1959, "Scattering of Impulsive Elastic Waves by a Rigid Cylinder," The Journal of the Acoustical Society of America, Vol. 31, pp. 1169-1175.

[282] Hudson, G.E., 1943, "Dispersion of Elastic Waves in Solids Circular Cylinders," Physical Review, Vol. 63, pp.46-51.

[283] Jones, O.E. and Ellis, A.T., 1963a, "Longitudinal Strain Pulse Propagation in Wide Rectangular Bars, Part I-Theoretical considerations," Journal of Applied Mechanics, Vol. 30, pp. 51-60.

[284] Jones, O.E. and Ellis, A.T., 1963b, "Longitudinal Strain Pulse Propagation in Wide Rectangular Bars, Part II-Experimental observations and comparisons with theory," Journal of Applied Mechanics, Vol. 30, pp. 61-69.

[285] Meitzler, A.H., 1965, "Backward-Wave Transmission of Stress Pulses in Elastic Cylinders and Plates," The Journal of the Acoustical Society of America, Vol. 38, pp. 835-842.

[286] Mindlin, R.D. and Fox, E.A., 1960, "Vibrations and Waves in Elastic Bars of Rectangular Cross Section," Journal of Applied Mechanics, Vol. 27, pp. 152-158.

[287] Pao, Y.-H., 1962, "The Dispersion of Flexural Waves in an Elastic, Circular Cylinder–Part 2," Journal of Applied Mechanics, Vol. 29, pp. 61-64.

[288] Shaw, E.A.G., 1956, "On the Resonant Vibrations of Thick Barium Titanate Disks," The Journal of the Acoustical Society of America, Vol. 28, pp. 38-50.

[289] Tolstoy, I., and Usdin, E., 1957, "Wave Propagation in Elastic Plates: Low and High Mode Dispersion," Journal of the Acoustical Society of America, Vol. 29, pp. 37-42.

[290] Torvik, P.J., 1967, "Reflection of Wave Trains in Semi-Infinite Plates," 1967, The Journal of the Acoustical Society of America, Vol. 41(2), pp. 346-353. .pp 239-245.

[291] White, R.M., 1958, "Elastic Wave Scattering at a Cylindrical Discontinuity in a Solid," The Journal of the Acoustical Society of America, Vol. 30, pp. 771-785.

Index